Toward a ZERO ENERGY Home

Toward a ZERO ENERGY Home

A COMPLETE GUIDE TO ENERGY SELF-SUFFICIENCY AT HOME

David Johnston
& Scott Gibson

Authors of *Green from the Ground Up*

D E D I C A T I O N

To Greg Franta and Gail Lindsey, inspirations and lights who have guided the sustainable building movement.

The Taunton Press, Inc.
63 South Main Street
PO Box 5506
Newtown, CT 06470-5506

e-mail: tp@taunton.com

Editor: Peter Chapman
Copy editor: Candace B. Levy
Indexer: Cathy Goddard
Cover design: Scott Santoro, Worksight
Interior design: Scott Santoro, Worksight
Layout: Noom Kittayarak and Scott Santoro, Worksight
Illustrator: Christopher Mills
Cover Photographer: Emily Minton Redfield

Library of Congress Cataloging-in-Publication Data

Johnston, David, 1950-
Toward a zero energy home : a complete guide to energy self-sufficiency at home / David Johnson & Scott Gibson.
p. cm.
Includes index.
ISBN 978-1-60085-143-8
1. Dwellings–Energy conservation. 2. Ecological houses. I. Gibson, Scott, 1951- II. Title.
TJ163.5.D86J64 2010
696–dc22

2010000576

Printed in the United States of America

10 9 8 7 6 5 4 3 2 1

Construction is inherently dangerous. Using hand or power tools improperly or ignoring safety practices can lead to permanent injury or even death. Don't try to perform operations you learn about here (or elsewhere), unless you're certain they are safe for you. If something about an operation doesn't feel right, don't do it. Look for another way. We want you to enjoy the remodeling process, so please keep safety foremost in your mind whenever you're working on your project.

ACKNOWLEDGMENTS

The authors gratefully acknowledge the generous support of the many people who made this book possible. While not exactly brand new, zero energy building is far from a finished science. The builders, architects, energy consultants, homeowners, and government researchers who spoke with us were instrumental in helping us understand both the promises and pitfalls of net zero design.

Thanks, first, to Deirdre Damron for her persistence, insight, relentless pursuit of photos and details, and coordination with the building professionals. Deirdre's research and editing have helped make this book what it is.

We are grateful to the owners of the zero energy homes that are featured in the case studies throughout the book. They willingly opened their doors and shared what they had learned. Thank you Jane Bindley, David Pill, and Hillary Maharam, Cador Pricejones, and Michael Kracauer.

Thanks to the builders, architects, and researchers who explained how zero energy houses work and provided valuable contacts in this expanding industry: Ben Southworth of Garland Mill Timberframes, Jeff Christian of the Oak Ridge National Laboratory, Paul Norton of the National Renewable Energy Laboratory, Danny Parker of the Florida Solar Energy Center®, Chris Benedict, Bruce Coldham, Paul Eldrenkamp of Byggmeister Design Build, Paul Fisette, David Barclay of the Northeast Sustainable Energy Association, Doug Parker (who is also the sustainability inspector for Boulder County, Colorado), and Jim Logan, who's a pioneering architect in zero energy homes. Thanks also to Eric Doub of Ecofutures™, who started the zero energy home movement in Colorado.

Zero energy design leans heavily on the expertise of energy consultants, engineers, wind and solar experts, and heating and cooling specialists. Our appreciation goes to Henry Gifford, Marc Rosenbaum, Andy Shapiro, Paul Gipe, and Adam Stenftenagel of Sustainably Built in Boulder for explaining the basics.

Thanks also to the many individuals at the National Renewable Energy Laboratory who went out of their way to help, including Sandy Butterfield, Cecile Warner, Jim Green, Vikki Kourkouliotis, and Mike Linenberger.

We're grateful for the support at The Taunton Press, especially from our editor, Peter Chapman, whose patience, good humor, and editing skills made this project possible.

Finally, we thank our families for their love and support during the inevitably long process of researching and writing a book. It's a journey we could not have made without them.

Contents

5

Every day, the sun provides enough energy to the Earth to meet the world's electricity needs for a year.

INTRODUCTION

The Case for Zero Energy Houses

The petroleum economy bared its teeth in 2008, and it wasn't pretty. In mid-July, the cost of a barrel of crude oil reached an all-time high of $147, a 50 percent increase in just 7 months and a threefold jump in 3 years. A few months later, as the world economy took a nosedive, prices dropped to less than $60 a barrel and gas prices dipped to nearly $2 a gallon. Heating oil customers in the Northeast who had locked in a winter's worth of fuel at more than $4.70 a gallon looked wistfully at a cash price of less than $2.20. It was the most recent upheaval in our fossil fuel economy, and it almost certainly won't be the last.

The cost of oil has a huge impact on every corner of the economy, in part because we continue to use so much of the stuff. Americans manage to burn more than 20 million barrels of petroleum products a day; about 12 million barrels of that is imported, making us the world's largest consumer. Until a few years ago, that didn't seem to matter. Oil was relatively cheap, and the Arab oil embargo of the 1970s was long forgotten. Now it matters a great deal as the developing world competes for a bigger share of this limited resource.

All of this affects how much we pay for energy. But the cost of fuel oil or gasoline looks like small potatoes in comparison with the environmental consequences of burning the enormous quantities of oil, natural gas, and coal we pull from the earth. Climatologists link an increase in atmospheric greenhouse gases to a steady rise in average global temperatures and a variety of climate changes, some of which may prove catastrophic. Carbon

Zero energy houses, like this one in Vermont, start with carefully detailed building envelopes. Controlling air leaks through exterior walls and roof and adding lots of insulation reduce the amount of energy needed for heating and cooling.

dioxide, a by-product of burning hydrocarbons, is an especially noticeable culprit. Glaciers are melting. Weather patterns are changing, bringing bigger, more frequent storms to some regions and droughts and high temperatures to others. High energy costs and a lack of potable water could make some parts of the globe very difficult places to live in the future. Worse, climate changes are occurring faster than scientists had predicted only a few years ago.

What does building houses have to do with any of this? A lot. According to the National Renewable Energy Laboratory, 40 percent of all primary energy consumed in the United States and 70 percent of the electricity produced by U.S. power plants goes directly to commercial and residential buildings. By some estimates, buildings are responsible for 48 percent of the carbon released into the atmosphere.

This is where sustainable building first found a toehold. Using less energy for heating and cooling makes houses less expensive to live in while reducing their environmental impact. Other fundamentals of green building help buildings last longer, give them healthier interiors, and helped reduce the natural resources needed to construct them. People got it. Green building has prospered. As we write this, green building is just about the only good news in the building industry.

The question is whether these gains will be good enough in a world where oil can cost $100 a barrel in January and $147 in July. There are other issues: How much will fossil fuels cost over the life of a house? How do we keep housing affordable on a monthly basis if utility bills approach the cost of the mortgage?

The Next Step: Net Zero

Net zero and near net zero homes take the goals of sustainable building one step further. But just what is a zero energy house?

Not too long ago, a house that used 60 percent less energy than one built to code was called a *near zero house*. That was quite an accomplishment. A house that was this energy efficient could be constructed mostly with conventional materials and techniques but with more attention to details, such as air sealing and insulation. Builders and architects got to thinking that going way beyond code might give them a market advantage.

Today, building standards are getting tougher, and there are many labels for super-efficiency. There are zero energy homes, net zero homes, carbon neutral homes, and off-the-grid homes. What are we talking about here?

The simple definition is that a net zero energy building produces as much energy as it uses on an annual basis. This includes energy for heating, cooling, and all the devices that plug into the wall. Net zero houses are typically connected to a local electric utility. They use the grid for storing excess electricity generated by photovoltaic panels or a wind turbine, banking electricity at times of plenty and drawing on the surplus when production falls.

Green-built houses aren't necessarily the same as net zero energy houses, although this lakeside home in New Hampshire manages to do both. It won a Leadership in Energy and Environmental Design accreditation and is also a net producer of electricity.

A house in a cold climate might need more energy than it makes during the winter but then makes up for it in summer when demand is lower and the photovoltaic system is running at full tilt. The opposite may be true in the south, where high humidity in the summer requires more electricity for air-conditioning during peak months. But on average, zero energy houses produce enough energy to offset the high-load months.

Most grid-tied homes are built where the local utility offers net metering. That means the utility will buy electricity at the same price it charges, but usually only until the net is zero. If houses produce more than that, the utility may buy it back. If so, it is often at the wholesale price, which can be as little as 1 or 2 cents per kilowatt hour when the retail price is 10 or 12 cents per kilowatt hour. That makes those excess electrons produced very expensive. In Germany, the government has imposed rates on utilities, forcing them to pay roughly 50 cents per kilowatt hour as an incentive to building owners to produce more electricity than they use.

Off-the-grid houses must provide all the electrical energy its occupants need, summer and winter. Other than relatively small battery banks, there is no place to store energy. The house is truly self-sufficient. For decades, a handful of builders around the country have experimented with off-the-grid approaches in different climates. They have taken very diverse paths to get to the end goal. Most often what makes the house self-reliant are changes in lifestyle for the families that live in them. Electricity goes on a budget. There is a fixed amount of energy available for any given day. If someone wants to take a hot shower, it has to be on a day with plenty of sun. If you want toast in the morning, maybe you can't use the hair dryer. Most Americans aren't willing to adjust their lifestyles that radically.

Retrofitting existing homes to near-net zero standards is especially challenging. This two-family house in Somerville, Massachusetts, got a new blanket of foam insulation on the outside to cut energy consumption sharply.

Houses also can be designed to produce enough energy to offset the embodied energy in all the building materials plus the energy required to build the house. This means the house must produce *more* energy than it uses on a yearly basis. Roughly 8 percent of a home's energy use is embodied energy from producing and transporting the building materials used in its construction. This is sometimes called regenerative architecture, and it has a deep ethical vein running through it.

A carbon-neutral home uses a different metric to determine how to get to zero. More than just zero energy, it must be zero carbon emissions all the way back to the power plant or manufacturing facility that made the building products in the first place. On average, getting electricity from a power plant to a house is at best 30 percent efficient. From a carbon-neutral standpoint, the electricity used from the grid has to be repaid with three times more site-generated electricity to break even. The same holds true for building materials. If the marble tile in the foyer is from Italy, the energy produced at the house has to be sufficient to make up for the embodied energy from extraction and transportation of the marble. The utility buy-back policy also dictates the financial context for this approach. Adherents to carbon-neutral houses are insistent on using only local materials and simple solutions to getting to zero energy. The more complex the house, the more diverse the sources of the materials and the more energy needs to be produced.

Size also comes into the net zero discussion. Some say a 10,000-sq.-ft. home can never be sustainable. It is just too big and energy and material intensive. How can a family of four need so much space when in developing countries 10 families would inhabit a house that size? Communities such as Marin County in California and Aspen in Colorado penalize houses that exceed a prescribed maximum square footage. The larger the house, the more energy efficient it must be until finally, at a certain size, code drives the design. Aspen allows a homeowner to buy his or her way out of this problem by putting money into a fund that pays for solar collectors on homes of police officers, firemen, and teachers. The result in net carbon may be the same in this Robin Hood approach to reducing carbon emissions.

But let's keep it simple. For the purposes of this book, *net zero* or *zero energy* means the house makes as much energy as it uses over the course of a year.

Building Houses a New Way

Regardless of the term we use, builders are realizing opportunity in this upside-down market by building homes that provide financial security for their customers. By minimizing utility bills or even creating the potential for the home to make money by selling energy to the utility at some point in the future, zero energy homes offer a new direction for housing in America. It is a win for the homeowner, for the planet in aggregate, and for a new generation of builders who will be able to construct houses that better meet future energy challenges.

This book includes a collection of case studies of zero energy or near zero energy houses from across the country, along with a description of the techniques and materials used to construct them. The builders, architects, remodeling contractors, and homeowners we spoke with aspire to reach that magical point of energy self-sufficiency. Some got closer than others.

Photovoltaic panels that generate electricity from sunlight are typically a key ingredient in energy self-sufficiency. Wind turbines sized for residential use are another option, but in either case, the relatively high cost of renewable energy puts a premium on reducing heating and cooling loads.

In any case, it is a journey of exploration into new approaches to the task of building meticulous, energy-efficient houses. Achieving the goal isn't as much the focus as is how they got there, how much it cost, what the options were, and what lessons builders learned along the way. When we started, we'd heard of a few net zero projects. At each interview we learned of a few more. We are starting to see a wave of interest from all parts of the country.

It's easy to spend a lot of money on a net zero house, but this near zero Habitat for Humanity home in Lenoir City, Tennessee, proves that high efficiency can come with a modest price tag.

Building a real net zero house is more than investing an arm and a leg in photovoltaic panels or buying a big wind generator. Reducing the *amount* of energy needed to heat and cool the house is the essential consideration, and that means a tight, well-insulated building envelope and more awareness on the part of homeowners about their energy use. Investing in energy-efficient appliances and lighting fixtures, eliminating phantom electrical loads, and orienting the house to take advantage of sunlight all cut the demand for electricity and fossil fuels.

But even taking all of these steps won't necessarily get a house all the way to the land of net zero. Much more realistic are near net zero homes. These houses also are designed to drastically reduce the amount of energy they use, which we'll explore in the following chapters, but they fall somewhat short of producing all the power they use. Many of the builders talked about the 80/20 rule; 80 percent of the load reduction can cost only 20 percent more than a standard house, but the last 20 percent can mean an additional 80 percent of the incremental cost. This is hardly a failure. What if all new houses in the United States used 90 percent less energy than what we use now? Even 80 percent less? The impact would be immense.

Because of the experimental nature of zero energy homes, most are custom made and many cost more that conventional construction.

There are scores of builders all over the United States and Canada who are constructing houses like these. A general public clamor for better energy performance has helped and so have a variety of public and private programs that promote zero energy construction. In Massachusetts, there is the governor's Zero Net Energy Buildings Task Force, announced at the Northeast Sustainable Energy Association's

2008 building energy conference. Architecture 2030, created by architect Edward Mazria, is pressing for changes in building design and construction that will make all buildings in the United States carbon neutral by 2030. The U.S. Department of Energy (DOE), the Canada Mortgage and Housing Corporation (CMHC), the California Energy Commission, and a variety of other public agencies have launched their own initiatives.

Building a net zero house can get expensive, but it's not necessarily so. In New York City, for example, architect Chris Benedict specializes in multifamily buildings that use a fraction of the energy that a conventional building of the same size would consume. Benedict is able to accomplish this without the use of any renewable energy systems and at a cost no greater than conventional construction. A net zero Habitat for Humanity® house near Denver, Colorado, was built for $116 a square foot (see the photo below). Near net zero houses built under the Habitat program in Tennessee have energy costs of about $1 per day. It's not just a game for the well heeled.

There is no single path toward energy self-sufficiency, nor are we arguing that building net zero houses will magically solve the world's energy or climate problems. But one house at a time, one neighborhood at a time, is how green building became mainstream. Building houses that are energy self-sufficient is completely within our capabilities—not at some distant point in time, but right now.

High-tech materials, like structural insulated panels and spray-in polyurethane foam insulation, help reduce energy loads—at a price. This net zero Habitat for Humanity project in Wheat Ridge, Colorado, was built with standard materials and techniques, suggesting that net zero construction is within reach of many American families.

1 The Building Envelope

The world of zero energy homes is all about rethinking the envelope, or outer layer, of the building, which includes the roof, walls, and foundation. The challenge is to build an envelope that is as airtight and well insulated as possible so that the energy that comes into the building stays within the envelope, and any inclement weather outside has little impact on the comfort inside.

By rethinking how we design and build the envelope to achieve zero energy, all other decisions like heating, ventilation, and air-conditioning (HVAC) and active solar become less expensive.

From a design standpoint, the guiding force in zero energy homes is solar orientation (specifically, Which direction is south?), which is discussed in detail in chapter 2. If we begin the design process thinking about how we can use the sun to its fullest potential, many other decisions become easier and less expensive. Zero energy homes take advantage of subtle energy flows, such as sunlight turning into heat and how to move that heat inside the building.

The building envelope requires very careful design and construction beyond what seemed adequate just a few years ago. Zero energy homes are often 50 percent or more energy efficient than their local energy codes require (see the sidebar on p. 12).

Building Science 101

Energy conservation in the building envelope is the initial focus for all the zero energy homes featured in this book. Reducing loads—the

A super-tight, well-insulated building envelope is the starting point for any zero energy home.

GOING BEYOND THE CODE

Energy codes, even the 2009 International Energy Conservation Code (IECC) update, are still behind the times. Codes are typically updated every 3 years and then have to be adopted by a local jurisdiction; thus by the time they are adopted they are often 5 years out of date. What assumptions do current energy codes rely on? Oil at $60 a barrel or oil at $147 a barrel?

The energy future we face is so unpredictable that the best way to anticipate it is to design homes that require as little fossil energy as possible. Even if "peak oil"—the point at which oil extraction has reached its maximum—is 10 years away, homes are built to last for at least 30 years and probably more. They will still have to be heated, cooled, and lit, so why build to a code that is based on yesterday's energy prices? With the uncertainty in energy prices, increased insulation and air sealing can be seen as either cheap insurance or the best investment you can make today. It is the lowest-hanging fruit to save energy that you can put in a building.

This zero energy home in Washington State was built with advanced framing 2×6 walls and spray-foam insulation.

To build well, it is critical to involve all the subcontractors in the design process from the beginning. Early collaboration helps prevent problems later in the project.

amount of fossil energy required to provide comfort year round–by half or more makes installing HVAC equipment much more affordable, as a smaller system is needed. Reducing loads requires getting all the building science right, which means understanding how building science principles work in a house: heat flow (thermodynamics), air flow, convection, the stack effect, controlling air with mechanical ventilation, and water flow (hydrodynamics). The key concept here is *durability,* so the house lasts as long as buildings built by our forebears. A durable home includes strategies for managing water in all forms (liquid, gas, solid), heat loss, heat gain, ultraviolet light, pests, and natural disasters.

Thermodynamics, or how heat moves (from *thermo*, meaning "heat," and *dynamics*, meaning "movement"), is something we are all familiar with, but the basics often escape us when we build. Heat moves from hot to cold–always–even though it may seem that it is the cold that makes us uncomfortable. That is why we call it "heat loss."

How heat moves

Heat is transferred in three ways: by conduction, convection, and radiation. *Conduction* is the way heat moves through a solid material. If you put a steel poker into a blazing fire, it isn't long before it gets too hot to handle. We measure conduction with R-values. The higher the R-value, the greater the resistance to heat flow. Insulation has a better resistance to heat flow than wood. Wood resists heat flow better than metal. Common sense tells us that a hot frying pan with a wood handle can be picked up with bare hands but a cast-iron frying pan needs a pot holder. What has this got to do with construction? Building with steel studs, for example, reduces the effective R-value of a wall assembly by 50 percent.

The heat from a fire is conducted by a metal frying pan.

The sun's radiation warms the earth.

Convection is how heat moves through a gas or liquid. Hot air rises, cold air falls. Putting a hand over the flames of a fire lets us know how rapidly the heat can be transferred. Convection is what makes a chimney work. In a house, dense cold air falls, pushing the hot air upward. The air cools and starts to decline again (this is known as a "convective loop"). In a conventional home, air hits a relatively colder window, falls to the floor pushing the warmer air up in the room, and causes drafts. It is what gives us cold feet in winter.

Radiation is how hot bodies transfer heat to colder bodies. We are typically most aware of radiation when standing in the sun on hot days, and we naturally move into the shade when we can. Radiation typically follows line of sight; if we can see a fire, we can feel the heat. If we go around a corner, the heat is not as palpable. In terms of a building, west-facing windows are heated by late afternoon sun, which increases the cooling load of the home in summer.

As simple as these laws of physics seem, they can get very complex inside a house when all three are at work simultaneously. By paying attention to conduction, convection, and radiation at every stage of envelope construction, building high-performance houses is more common sense than something exotic.

Air movement

Air leakage is a major problem in most houses. A typical house has 2,000 linear ft. of cracks and gaps that allow air in and out, which can represent up to 50 percent of the heat loss in

THE HIGH-WIND EFFECT

High wind can make air and moisture control more challenging. In Colorado, houses are subject to winds of over 100 mph every fall and winter. Wind-driven air flows create high pressure on the windward side of the house and negative pressure on the leeward side. This accelerates heat loss and can reduce the effectiveness of insulation that doesn't block air flow horizontally through the wall. Wind flowing through fiberglass insulation at 15 mph, for example, can reduce the R-value from R-15 to R-4.

a building. Often, the older the house, the more air leakage it has. Wherever we join building elements together—bottom plates to subfloor, studs to stud corners, top plates to trusses—there is the potential for air movement through the shell of the building. The designer must be able to identify where this air flow is and how it should be sealed. The air barrier is where outside air is separated from inside air, and there needs to be an air-impermeable separation between the two zones. The air barrier can be toward the inside of the house or toward the outside, depending on climate and selection of building materials.

Not only does air leakage drain energy from the home but it also often carries moisture into wall cavities or attics, which leads to mold and reduces the lifespan of the house.

The tighter the house, the more important it is to deal with both air and moisture movement inside the building. Air and moisture leakage can be blocked by air barriers and vapor retarders. An air barrier stops the movement of air into and through the cavities of the building. A vapor retarder slows the migration of moisture into the walls and ceilings. A vapor retarder may or may not be the same material as the air barrier. For example, closed-cell spray urethane insulation (see p. 37) in rafter spaces provides both an air barrier and a vapor barrier. A wall insulated with fiberglass and wrapped with taped rigid foam insulation has an air barrier on the outside of the sheathing but a vapor retarder on the inside of the wall provided by the drywall and latex paint.

To compound the complexity, as warm air rises, it flows through any vertical holes into the attic, a phenomenon known as the *chimney* or *stack effect*. If that warm air contains

Air sealing is a key element in creating a tight building envelope. Using expansion foam is an easy and effective way to seal air gaps.

water vapor, it will find its way into the attic and condense on the colder roof sheathing, sometimes actually raining on the ceiling insulation below. Air sealing needs to start during construction so that all holes between floors are caulked and sealed before the insulation is installed. Electrical, plumbing, and HVAC flues are likely suspects in the first stages of air sealing. Conducting a blower-door test before the drywall is installed (see p. 39) can help find remaining major air leaks.

Water intrusion

Air movement and water movement are interconnected and must be considered as the building is being designed. For decades, the building industry has separated the two, which has led to numerous building failures. It has always been the job of builders to keep water out of the house, and typically we depend on the roofer to make sure the house doesn't leak. But that is dealing only with water, not water vapor or, in extreme conditions, ice. We have to take all three into account to ensure the long life and durability of the house.

Similar to heat, water always flows from higher concentrations to lower concentrations and from warm to cold. If it is more humid inside a house than outside, the pressure will move water vapor through the envelope to the outside. In hot, humid climates with high cooling loads, the pressure will be reversed. This is where air movement and water movement come together. The water vapor is carried in the air, and wherever air can move so can water. To keep moisture out of the wall cavities, it is critical to define the location of the air barrier.

Moisture is the enemy of the building world. A good envelope is designed to shed moisture before it causes problems.

Water vapor control is very climate specific and is determined by the combination of various products (paint, drywall, insulation type, sheathing, drainage plane, and siding). Because air flow is dictated by both temperature and pressure, and moisture is driven by humidity (water moves from wet to dry), vapor retarders in cold climates should be toward the inside of the home. In hot, humid climates the vapor pressure is from the outside in, thus the vapor retarder should be toward the outside. In either case, the wall cavity, if it is able to absorb moisture, wants to dry toward the opposite side of the vapor retarder. Complicated enough?

Water movement from the outside of a house to the inside is one of the major causes of building failure. While a house might be sealed well when originally constructed, over time, building elements can deteriorate. For

Housewrap keeps vapor out. Flashing is used to keep water from damaging or rotting the wall system.

Mold grows in moist environments, where it feeds on cellulose, such as the paper backing on drywall. Not only is mold unsightly, but it also can cause serious health problems.

example, adhesive flashing around windows and doors may work well for the first few years, but eventually adhesion may fail, creating pathways for water to creep around the flashing and into the framing. Mechanical flashing, or ensuring that each layer of housewrap correctly laps over the previous layer, is especially important above doors and windows. Flashing and housewrap should always work together and be lapped from the bottom up, just as roof shingles always lap over each other. Manufacturers of housewraps and window flashings provide detailed installation information, yet most windows are still installed incorrectly.

Learning from the Passive House Movement

The idea of a superinsulated "Passive House" is a well-thought-out approach to building a zero energy home. The *Passivehaus* concept was developed in Germany in 1990 based on passive solar research conducted in the 1970s by the U.S. Department of Energy (see chapter 2). Good passive solar design can reduce the thermal load of a building by 90 percent, primarily through super-insulation, an airtight envelope, good windows, and heat-recovery ventilation. Most of the thermal comfort is provided by passive solar heat, the waste heat from electrical equipment, and people's daily indoor activities. The rest of the heat is often provided by small electric-resistance heaters. Today there are thousands of Passive Houses across Europe.

Passive House principles

When considering building a zero energy home, Passive House standards actually provide a better place to start than local energy codes. They provide an almost ideal set of criteria against which to gauge how close to zero energy you want your house to be.

Using modeling software

Modeling a zero energy house is a critical first step and makes up about half the work in getting the results in actual performance. It is a lot easier to correct mistakes on a computer than it is on the job site. Software modeling allows the design team to test a variety of assumptions concerning how much insulation is needed, how passive solar will be used to reduce heating requirements, and how much active solar to add at the end to meet the rest of the home's need for energy.

The Passive House Planning Package (PHPP) software is recommended by the Passive House Institute since it is designed for very low energy envelopes. The software conducts energy calculations for the entire house, taking into account orientation, house size, location of windows and overhangs, insulation levels, airtightness, and natural and mechanical ventilation. The PHPP will also factor in solar water heating.

THE PASSIVE HOUSE MOVEMENT IN AMERICA

The Passive House movement has started to gain a foothold in the United States. There are Passive House groups in major cities across the country, with the epicenter of the U.S. Passive House movement in Urbana, Illinois, at the Passive House Institute U.S. The first Passive House built in Urbana in 2003 has 12 in. of fiberglass wall insulation plus 4 in. of polystyrene sheathing on the exterior (R-60+) and roof insulation of 16 in. of blown fiberglass (R-64). About 5 kilowatts of solar photovoltaics were required to make it a net zero house.

Constructed with insulated concrete forms and double-framed stud walls, this home in Isabella, Minnesota, is built to Passive House standards.

GETTING TO ZERO

The single best way to keep energy costs down and performance up is to tighten up the building envelope. That entails using the appropriate energy modeling software during the design phase. The designers of each of the homes featured in this book used energy modeling to determine where to invest money to get to zero energy. Climate plays the pivotal role in starting the process. Each climate requires a different solution, as can be seen in the case studies. Remember that different software delivers different projections in energy performance. Some programs treat the south wall of the building the same as the other orientations. Some model the passive solar performance and help size thermal mass. Others incorporate solar thermal and photovoltaics. It is important to use the software that best suits your design requirements.

Many other home energy modeling software programs are also available (see "Getting to Zero" above and Appendix 1 on p. 242). When choosing which software to use, it is important to consider how well the software handles the passive solar component and other solar features. Some software does not take orientation and passive solar gain into account, which can lead to an oversize HVAC system. A final word of caution: Software is a useful tool, but it isn't perfect and can yield erroneous results relative to actual home usage.

Superinsulate the envelope

The correct amount of insulation is one of the key aspects to address when tightening up the building envelope. The National Renewable Energy Laboratory (NREL) advocates a simple formula when it comes to insulation: 30-40-50. In colder climates, zero energy homes start with R-30 for floors, R-40 for walls, and R-50 for ceilings/roofs. Farther north, builders are using even higher numbers. Your climate will dictate the absolute R-values necessary to reduce the thermal load to close to zero. Most designers of zero energy homes are increasing insulation by 50 percent or more than their local code requires.

Eliminate thermal bridges

The greater the insulation values designed into the home, the more important it is to eliminate thermal bridging. In a typical 2×4 wall with studs at 16 in. on center (o.c.), the heat loss through the studs can reduce the overall wall performance by 15 percent. Over 20 percent of the wall area can actually be wood (at about R-1 per inch). If steel studs are used, the thermal bridging can reduce the effective R-value of the wall by nearly 50 percent. With thicker studs, the thermal bridging numbers are lower, but the losses are still unacceptable. This is why conventional framing should be covered with rigid foam on the exterior to reduce the thermal bridging losses.

AN OVERVIEW OF PASSIVE HOUSE PRINCIPLES

The principles of the Passive House stress the importance of a super-tight, well-insulated thermal envelope that allows the use of very low amounts of fossil energy to heat the home. According to Passive House standards, the heating requirement cannot exceed 1.4 kilowatts per square foot (kW/sq. ft.) per year. A typical house built to code uses 14kW/sq. ft. to 18kW/sq. ft. per year. Passive Houses mostly rely on the "free" heat given off by the occupants, appliances, and equipment to heat the home.

The Passive House standards are as follows:

- **Use the *Passive House Planning Package* (PHPP) software to model the house.**
- **Superinsulate.**
- **Eliminate thermal bridges.**
- **Make the house airtight.**
- **Specify heat-recovery ventilation.**
- **Optimize passive solar design.**
- **Specify high-performance windows and doors.**
- **Use internal heat gain (people, appliances, electronic equipment).**

Exterior 2-in. rigid closed-cell foam insulation eliminates the problem of thermal bridging.

Design for airtight construction

In the world of contemporary building science, airtight construction is a critical component. In the past, we thought that letting air into the house for ventilation was a good thing. We now know that unintentional air infiltration is not only uncomfortable but also often the cause of mold and moisture problems. Barbara Harwood, a builder from Dallas who has built many low energy homes, puts it this way: "There is no such thing as a building that is too tight, just one that is underventilated." In a typical new home today, inside air is exchanged with outside air about once every two hours. This provides plenty of fresh air for occupants but has an energy penalty: As much as 50 percent of the heating or cooling load of the house is lost due to this infiltration/exfiltration process (unhappy customers know it as "drafts").

The object in a zero energy design is to make the house so tight that it exchanges air only once every 10 hours or so. We make up the fresh air with mechanical ventilation and heat-recovery ventilators (see chapter 4) and, in so doing, control airflow. We can place air intakes where we want them and distribute the air in the house where it's needed. Without mechanical ventilation, 0.35 natural air changes per hour (NACH) is as low as most codes will allow you to build. That is about the limit of a conventional new, fiberglass-insulated house.

To get to airtight construction, the building needs to be wrapped with an air-sealing barrier or insulated with spray foam. Building products such as structural insulated panels (SIPs) can also create a very airtight structure. The critical areas to examine for air leaks are where there are penetrations, such as doors, windows, vents, and flues, and where building elements come together. The connectivity from floor to floor is also key, particularly between the top floor and the attic.

MAKING HOUSES AIRTIGHT

With so much potential for energy loss with a drafty building envelope, net zero builders spend a lot of time sealing potential air leaks. There are a variety of techniques that can help, but in the end the builder gauges success by calculating how long it takes for a complete exchange of inside and outside air. This is called *air changes per hour* (ACH).

In conventional construction, the ACH might be 0.5, meaning that it takes roughly 2 hours for a complete air exchange. Net zero builders must do much better than that. One tool at their disposal is a blower-door test, a specialized piece of equipment that fits in a doorway and depressurizes the house with a fan. The standard is to run the test at 50 pascals of pressure and measure the results, which are usually noted as "ACH50."

The relationship between natural ACH (or NACH) and ACH50 can be calculated. For a complete explanation of how that works, see Appendix 2 on p. 243.

AIRTIGHT CONSTRUCTION TECHNIQUES

Airtight construction is achieved with a variety of materials, including caulks, foam sealants, membranes like housewraps, gaskets, and high-temperature sealing around flues. During design, consider eliminating as many penetrations as possible by using Studor® air-admittance valves (AAVs) to vent the plumbing drainage system of the house. Studor AAVs provide ventilation without the need for roof penetrations and conventional vent piping.

Bathroom exhaust fans should be run through the heat-recovery ventilator (HRV) instead of using a separate fan. Use a condensing dryer instead of venting to the outside. Rethink incorporating decorative fireplaces as these are among the largest sources of air leakage in a home. Blower-door testing is essential to identify any air leaks that may have been overlooked before the drywall is installed, such as around a window frame that was improperly foamed.

Specify heat-recovery ventilation

When building an airtight envelope it is necessary to provide mechanical ventilation. Climate extremes will determine what type of ventilation is required. In mild climates with few heating and cooling degree days, passive fresh air intakes linked to an exhaust fan may suffice. These are through-the-wall vents with flaps on the inside that work by drawing in air when an exhaust fan is turned on. Otherwise, the flap seals and keeps air pressure in the house neutral.

In more extreme climates, whole-house mechanical ventilation is provided by a heat-recovery ventilator (HRV) in cold, dry climates or an energy-recovery ventilator (ERV) in hot, humid climates. These devices are basically air-to-air heat exchangers that in winter pull the heat from the exhaust air stream and transfer it to incoming air. The process is reversed in summer. They use very little electricity and can be up to 90 percent efficient in the heat transfer process. An ERV also captures the humidity in the air and transfers it to the opposite air stream, depending on season. In hot, humid climates one of the most important jobs of the air-conditioning system is to dehumidify the air. This is important to reduce the risk of mold in the house.

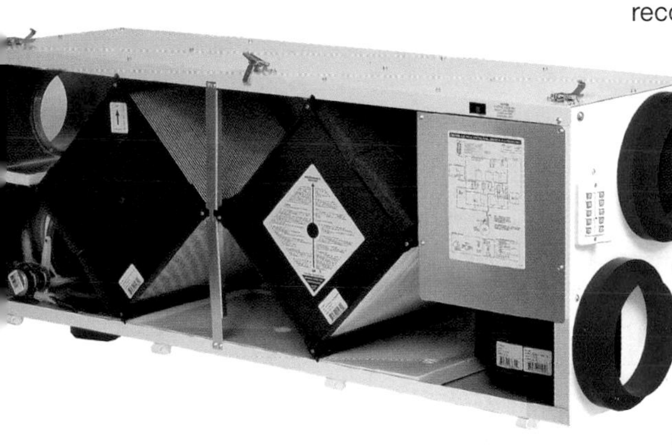

A heat-recovery ventilator helps reduce energy loss by conditioning fresh air while exhausting stale air.

This advanced window frame uses two Heat Mirror® films to reflect heat back and help control the interior temperature of the home.

Both systems can be operated using timers so that fresh air is continuously introduced into the house.

Optimize passive solar design

Passive solar heating through south-facing windows is the simplest and most cost-effective source of house conditioning. Once the insulation is up to Passive House standards, it takes very little passive solar gain to keep the house comfortable whenever the sun is out in the winter. (For more on passive solar design, see chapter 2.)

Install high-performance doors and windows

In the 1970s, low-e windows that improved the U-value of a typical double-pane window from U-0.50 to roughly U-0.35 were state of the art. By coating the interior side of each pane of glass, low-e windows reflect heat back toward the warmer side of the glass. The technology has improved dramatically over the last three decades with low-e squared and other innovations. Canadian manufacturers and now several American window manufacturers are producing excellent very low U-value windows.

A U-value is a measure of how much energy is lost through a window unit. It is the inverse of the R-value (see "What Does the Window Sticker Mean?" on the facing page). A whole window U-value of 0.33 has an R-value of 3, which used to be considered a high-performance window. Now U-values as low as 0.07 (R-value of 14) are possible by using insulated frames, multiple panes of glass, or plastic films with low-e coatings and low-conductivity spacers between the panes of glass. High-efficiency windows open up significant architectural possibilities and enhance the overall performance of the envelope. The technology is also being incorporated into glass doors and, ironically, glass doors can now have higher R-values than solid doors.

WHAT DOES THE WINDOW STICKER MEAN?

World's Best Window Co.

Millennium 2000+
Vinyl-Clad Wood Frame
Double Glazing • Argon Fill • Low E
Product Type: **Vertical Slider**

ENERGY PERFORMANCE RATINGS

U-Factor (U.S./I-P)	Solar Heat Gain Coefficient
0.35	**0.32**

ADDITIONAL PERFORMANCE RATINGS

Visible Transmittance	Air Leakage (U.S./I-P)
0.51	**0.2**
Condensation Resistance **51**	—

Manufacturer stipulates that these ratings conform to applicable NFRC procedures for determining whole product performance. NFRC ratings are determined for a fixed set of environmental conditions and a specific product size. NFRC does not recommend any product and does not warrant the suitability of any product for any specific use. Consult manufacturer's literature for other product performance information.
www.nfrc.org

The National Fenestration Rating Council (NFRC) energy performance label allows you to compare one window with another in terms of resisting heat buildup in summer, warming your house in winter, keeping out wind, and resisting condensation.

U-factor

The U-factor measures how well a window prevents heat from escaping. The rate of heat loss is indicated in terms of the U-factor (U-value) of a window assembly. A U-factor is the inverse of an R-value. If you divide the U-factor into 1 you get the equivalent R-value. U-factor ratings generally fall between 0.07 and 1.20. The lower the U-value (the higher the R-value), the greater a window's resistance to heat flow and the better its insulating value. For example, a U-.33 = R-3 (typical for low-e windows); a U-.50 = R-2 (typical for double-glazed windows).

Solar heat gain coefficient

The solar heat gain coefficient (SHGC) measures how well glass blocks heat in sunlight. Heat is carried by the infrared spectrum of sunlight. The SHGC is the fraction of solar radiation admitted through a window (both directly transmitted and absorbed) and subsequently released inward. Another way of thinking about it is how much heat gain is blocked by the window. SHGC is expressed as a number between 0 and 1. The lower a window's solar heat gain coefficient, the less solar heat it transmits in the house. The lower the number, the better for east and west windows; the higher the number, the better for south windows.

Visible transmittance

Visible transmittance (VT) measures how much light comes through a window. It is an optical property that indicates the amount of visible light transmitted. VT is expressed as a number between 0 and 1. The higher the VT, the more visible light is transmitted. This is primarily important if retrofitting windows on the same side of a house with existing windows. They will look different from both inside and outside if the VT number is different. If all windows have the same VT, your eyes can't tell the difference until you open a window. Think of this as sunglasses for your windows.

Air leakage

Air leakage (AL) is indicated by an air leakage rating expressed as the equivalent cubic feet of air passing through a square foot of window area (cfm/sq. ft.). Heat loss and gain occur by infiltration through cracks in the window assembly. The lower the AL, the less air will pass through cracks in the window assembly. This number is important to the airtightness of the building.

Condensation resistance

Condensation resistance (CR) measures the ability of a window to resist the formation of condensation on the interior surface. The higher the CR rating, the better the window is at resisting condensation formation. Although this rating cannot predict condensation, it can provide a credible method of comparing the potential of various windows for condensation formation. CR is expressed as a number between 0 and 100. Each surface of glazing raises the CR of the window. Super-windows have a very high CR.

GETTING TO ZERO

There are many innovative ways to frame a zero energy home. In all cases, airtight construction and limiting thermal bridging losses drive framing decisions.

Managing the load

In most climates, the thermal design of the house is driven by the envelope. When a house is superinsulated, the internal heat generated by people, appliances, and lighting can contribute to winter comfort but can also be a source of unwanted heat in summer. Electrical loads inside can be reduced by technical solutions (for example, by using LED lighting rather than incandescent bulbs) and also by examining lifestyle choices.

In all cases, designing a zero energy home is much more about load management than it is about increasing the amount of photovoltaics one puts on the roof. To maintain affordability, both in initial costs and in operating costs, the key is to minimize the requirement for fossil fuels or electricity generated offsite.

Specific electrical circuits should be designed for optimal use and load reduction. For example, in entertainment rooms or offices with many plug-in devices, the circuit can be designed to shut off electricity to all devices at once. Experience has shown that most people just don't turn off individual plug strips that power entertainment centers with TVs, cable boxes, and all the various components that feed them. One switch can simplify both lifestyle and energy conservation.

The Diversity of Framing Practices

Conventional house framing has changed very little over the last few decades. After the oil embargo in the 1970s, some states like California established an energy code that required houses to be more energy efficient than they had been previously. Wall framing moved from 2×4s at 16 in. o.c. to 2×6s, which enabled contractors to fit R-19 insulation in the walls rather than R-11. In some parts of the country, there is still no energy code so the "cheaper the better" attitude prevails, which usually means the house sacrifices energy efficiency to upfront costs.

Advanced framing

Advanced framing (also known as optimal value engineering, or OVE) is based on studies conducted by the National Association of Home Builders Research Center in the late 1970s. Advanced framing techniques reduce dimensional lumber use by placing framing only where it is truly necessary. Practices include building walls at 24 in. o.c. rather than 16 in. o.c. and sizing headers to support true loads rather than wasting lumber where headers are not needed. The reduction in lumber leaves more room for insulation. Wherever there is framing there is no insulation. Framing materials form thermal bridges through the insulation, resulting in cold spots at every stud, especially in corners. At best, this causes discomfort and higher energy bills; at worst, it creates conditions in which condensation can occur and mold can grow.

ADVANCED FRAMING

Conventional Framing	Advanced Framing
2×4 16 in. o.c.	2×6 24 in. o.c.
Random load spacing: 16 in., 19.2 in., 24 in. o.c.	Stacked loads at 24 in. o.c.
4- to 5-stud corners	2- to 3-stud corners
Double 2×8 headers	Headers engineered for loads and insulated
3-stud partition wall intersections	Ladder blocking at intersections
2×12 rafters	2×4 trusses

This home was built using advanced framing techniques and insulated with open-cell urethane foam. The studs are 24 in. o.c., and nonbearing walls do not have headers.

The use of 2×6 advanced framing reduces thermal bridging and requires less lumber.

Production and custom builders who have converted to the advanced framing approach have reported at least a 20 percent reduction in materials and 15 percent reduction in framing labor, but thermal bridging continues to be a problem. Thermal bridging can be almost eliminated by wrapping 1 in. to 2 in. of rigid closed-cell foam (R-5 to R-10) around the building, under the housewrap. The foam can take an R-19 wall to R-29 and will reduce infiltration and create a quieter home. Perhaps one of the greatest advantages is that in cold climates the dew point (the temperature at which airborne moisture condenses) is moved outside the envelope to somewhere in the exterior foam, which dramatically reduces the potential for mold to grow inside the wall cavity.

Adding 1 in. of foam to the exterior affects openings like doors and windows. An inch of rigid foam and stucco or cementitious siding can be accommodated in the design by using brick mold casing, a 5/4 material that captures the ends of the siding and insulation; 5/4 milled wood can also work for wider cas-

This roof is wrapped with Titanium™ UDL roofing underlayment, which, unlike asphalt felt paper, is 100 percent inert to moisture absorption. Using this recyclable underlayment eliminates the incidence of water breach and the risk of mold.

ing. If the insulation is thicker, then different approaches need to be taken. In conjunction with exterior foam and over-the-house wrap, builders like Dennis Allen in Santa Barbara use vertical furring strips, nailed to the studs to attach siding and trim.

Double-frame walls

Precut 2×6s are common framing materials, but a number of builders use 2×8s (either solid sawn boards or laminated veneer lumber) to provide a thicker wall cavity for insulation. The downside, of course, is that the wider framing lumber adds to the cost.

Habitat for Humanity in Denver has come up with a solution that is both more affordable and easy to implement. Working with the National Renewable Energy Laboratory in Golden, Colorado, they devised a double 2×4 wall with a $3\frac{1}{2}$-in. gap between the walls (see the right photo below). This approach eliminates thermal bridging entirely and creates a $10\frac{1}{2}$-in. wall to fill with insulation. It requires using $\frac{3}{4}$-in. pressure-treated plywood for the top and bottom plates. The studs are staggered so the insulation fills the cavities completely. While this is a great solution, it's important to note that building two walls is double the work. Habitat had the advantage of extra labor, but this may not work as well for conventional builders who prefer on-site labor to be as efficient as possible.

Siding installed over furring strips creates a drainage plane to shed moisture without causing rot or mold.

Using double 2×4 wall construction with a gap for insulation is one of the cheapest and easiest ways to build an extra-thick wall.

Wood I-joist studs

Another way to build thicker walls is to use wood I-joists for studs. Pressure-treated plywood or OSB is typically used for the top and bottom plates, and blocking is often used at the 4-ft. level to strengthen the lateral flexibility of the joist studs. (Blocking is particularly important if spray foam is used for insulation because the foam can expand and warp the I-joists laterally.) Another advantage of I-joists is that the thinner section through the OSB web slows heat flow so there is less thermal bridging through to the exterior. I-joists also make the wall assembly lighter to erect than 2×6 framing.

I-joist wall framing creates a thick wall with low thermal bridging and lots of room for insulation. I-joists are also much straighter than conventional studs, which makes hanging drywall easier.

Structural insulated panels, which are usually installed with a crane, create a very tight building envelope.

Structural insulated panels

Using structural insulated panels (SIPs) is probably the easiest way to make thicker walls. SIPs come in any thickness needed and can be cut to any size and configuration. An 8-in. SIP filled with expanded polystyrene has an R-value of 32, whereas a 6-in. panel has an R-value of 24; for polyurethane foam SIPs, a 4-in. panel has an R-value of 24 and a 6-in. panel has an R-value of 38. Door and window openings are precut and typically fit like a glove. When the panels are cut perfectly, the erection process for a whole house can take just a few days (versus weeks for stick framing). However, SIPs are not always cut perfectly and precious time can be lost if one or more panels have to be modified.

In addition to high R-values, SIPs have the added benefits of no wall cavity where moisture can condense and create mold problems. And with no wall cavity, there is no air movement through the wall, so air sealing is easier.

If water does get to the panel, the "structural" in a SIP goes away. If SIPs get wet and stay wet, the OSB can start to delaminate and swell so you end up with structural foam. SIPs have to be caulked and sealed between panels during installation and then again when the structure is up. The exterior also needs to be carefully waterproofed. Housewrap and drainage planes are critical to ensure that no moisture gets onto the exterior OSB surface. Most SIP failures can been traced back to installation errors or moisture-sealing issues.

MANAGING MOLD

A wall cavity with insulation that allows air to move through it is one of the biggest problems with mold management. It's best to stop air flow through the building materials and direct all of it through the mechanical ventilation. By using materials that don't allow air movement, much of the mold growth can be prevented. Structural Insulated Panels, urethane foam (both open and closed cell), and tightly sealed buildings with advanced infiltration reduction measures can all protect the home from mold.

Attic and Roofing Systems

The attic needs to function as a system, which means that all the pieces need to work together to prevent moisture problems. This system starts at the drywall on the ceiling of the top floor and continues to the roofing material on the roof. Felt paper and roofing function to keep water out of the building, and the attic buffers indoor temperatures from outside temperatures. Attic ventilation evacuates moisture out of the enclosed space and keeps the underside of the shingles cooler, while providing space for thermal insulation.

The whole-systems approach initiates new ways of thinking about this subsystem of the house, particularly if ducts are run in the attic. The required first step is to define where the air barrier lies. If it is in the ceiling of the space below, it must be continuously unbroken. This is difficult when vents, ducts, recessed-can fixtures, flues, chimneys, and other penetrations perforate the air barrier.

Each must be sealed perfectly before insulation is installed above. If air leakage is coming through the air barrier, moisture can get into the insulation or, worse, into the attic space and condense on the underside of the sheathing, causing potential mold problems.

Ductwork in the attic causes a whole host of unintended consequences. In conventional housing, the air handler is often placed in the attic. In hot climates like Phoenix, this is standard procedure. In summer, air temperatures in the attic can reach 160°F. Cold air flows through flex duct in the attic that perhaps has R-6 duct insulation around it. The coldest air flows through the hottest space with the least insulation separating the two. From a systems point of view this makes very little sense.

This scenario calls for a semi-conditioned attic, which is accomplished by putting the air barrier on the underside of the roof sheathing. One way to create this air barrier is to spray closed-cell urethane foam (see p. 37) from the intersection of the walls with the roof framing to the ridge. The foam serves several purposes. It attaches the roof to the walls, providing structural support in high-wind, tornado, or hurricane-prone areas. The Federal Emergency Management Agency (FEMA) has found that the roof attachment is four times stronger with foam than with nailed trusses. It also provides

This vaulted ceiling is insulated with open-cell urethane foam insulation, which creates a strong roof system.

A semi-conditioned attic with insulated air ducts helps keep conditioned air from being affected by the temperature swings of a typical attic.

a complete air seal at the rafter level. Foam reduces the temperature of the attic space, providing a better place to run ductwork. It can also be used as a roofing material itself on commercial buildings and can divert water out of the attic if the roof fails. On occasion, when required by an inspector, a foam or plastic baffle can be placed under the roof decking connecting soffit vents to a roof ridge vent, which provides ventilation under the shingles. The spray foam is then applied under the baffle, which places ventilation where it belongs, above the insulation but below the shingles, essentially outside the envelope.

Another way to create a semi-conditioned attic is to use SIPs for the roof decking. This doesn't provide the positive air sealing that spray foam does, but it can move the dew point outside of the envelope. In either case, the attic/roof becomes a more functional space both thermally and in terms of reducing mold potential. Some ventilation must be provided. Typically with a forced-air system or with an HRV, a small register is placed in the attic with a return at the opposite end to provide air flow to make sure the attic space stays dry.

SPRAY POLYURETHANE FOAM PERFORMANCE

Polyurethane foam is most effective at preventing heat loss with only a couple of inches of insulation. More foam is needed to keep the surface temperature of the foam closer to the inside temperature of the building.

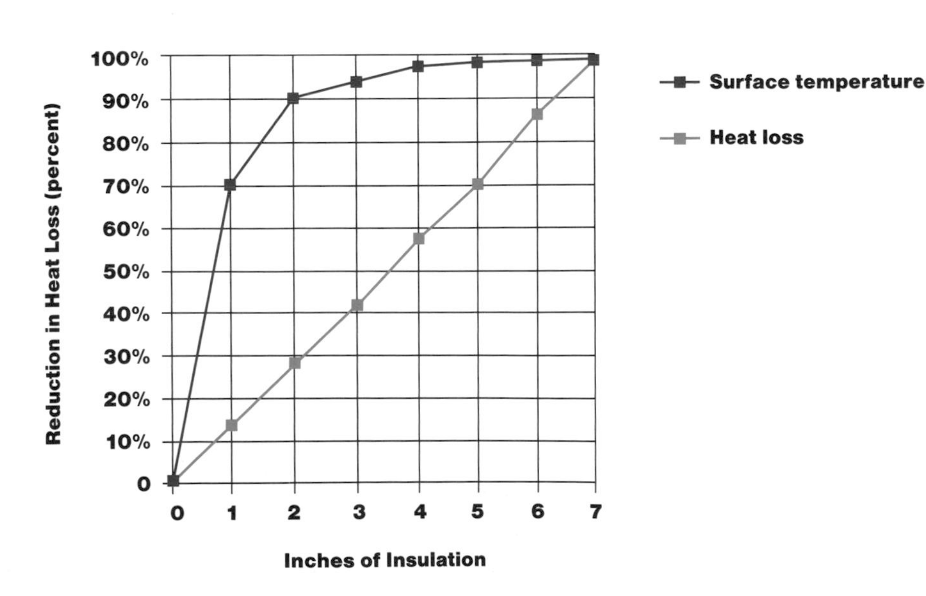

Basements and crawl spaces

The place where the building meets the earth is an area where many problems can occur. Water causes the majority of building failures, and homes need to be designed and built to keep water problems at bay. Careful attention to building science is critical in these areas.

In the building envelope, the main issues are air pressure and thermal performance. In the basement or crawl space, water is added to the mix. Hydrostatic pressure from water in the soil can force water into the basement. Capillary action can draw water through the concrete footings or slab into the interior space. Moisture that gets into the basement can be drawn into the house through air flow or leaky ductwork, adding to moisture problems in the spaces above.

The best way to deal with underground concrete is to isolate it from the soil. To help avert future water issues, you need to use a good waterproofing barrier (not just dampproofing). This barrier can be a rubber-based material that is painted on or an elastomeric membrane (such as is used on commercial roofs) attached to the concrete walls.

Thermal protection is best accomplished with closed-cell extruded polystyrene board placed on the outside of the concrete. Depending on climate, a 2-in. to 4-in. layer completely wrapping the exterior of the foundation keeps the foundation mass on the inside of the house and the dew point outside the envelope. If the basement is to be finished, then moisture is reduced and condensation is virtually eliminated. If it is slab on grade, 2 in. of rigid insulation should be used inside the forms and attached to the perimeter of the slab. If radiant heating is embedded in the concrete, the slab must be insulated with at least 2 in. of foam under the entire slab.

Rigid foam placed below a concrete slab reduces thermal bridging. An insulated slab increases the comfort in the basement and saves on heating costs.

Insulation Choices and Strategies

Insulation levels throughout the house are where zero energy buildings diverge from conventional energy code requirements most dramatically. To reduce net energy use to zero, houses need significantly higher insulation levels than conventional buildings that rely on abundant and inexpensive fossil fuels.

The obvious question is how much insulation is appropriate? The answer varies greatly depending on climate, but it's a safe bet that current government recommendations are inadequate. In Canada, some zero energy designers are installing attic insulation rated to R-65, which is 50 percent higher than current U.S. recommended minimums, or even to R-100. Canada, of course, experiences some brutally cold conditions, but even in the lower 48, we are seeing R-60 ceiling and R-30 wall insulation being installed. The real issue is how to get the heating and cooling loads so low that a homeowner could heat the house with the refrigerator, a computer or two, and the dogs. Low heating and cooling loads are the magic bullet that allows zero energy houses to be cost-effective in any climate.

Insulation types and location determine how well the home will perform through the diversity of climate conditions and future challenges. For example, using spray foam to air-seal the house makes subsequent insulation strategies easier than trying to air-seal the house with just fiberglass. Fiberglass insulation that is installed over a couple of inches of spray foam is one way to make the house airtight and still affordable. However, it's important to get the right amount of foam on the outside walls based on your climate, or major moisture problems could result.

GETTING TO ZERO

It is not enough just to insulate to meet minimum code standards. With Energy Star® requirements or Leadership in Energy and Environmental Design (LEED) standards, insulation is upgraded in conjunction with windows and mechanical equipment to bring the whole house to 15 percent better than local building codes. Although this is an improvement over code, zero energy homes have much higher targets.

Fiberglass

Conventional installation of fiberglass doesn't yield the performance required for a zero energy house. Typically, fiberglass is stuffed into wall cavities without regard for wiring, plumb-

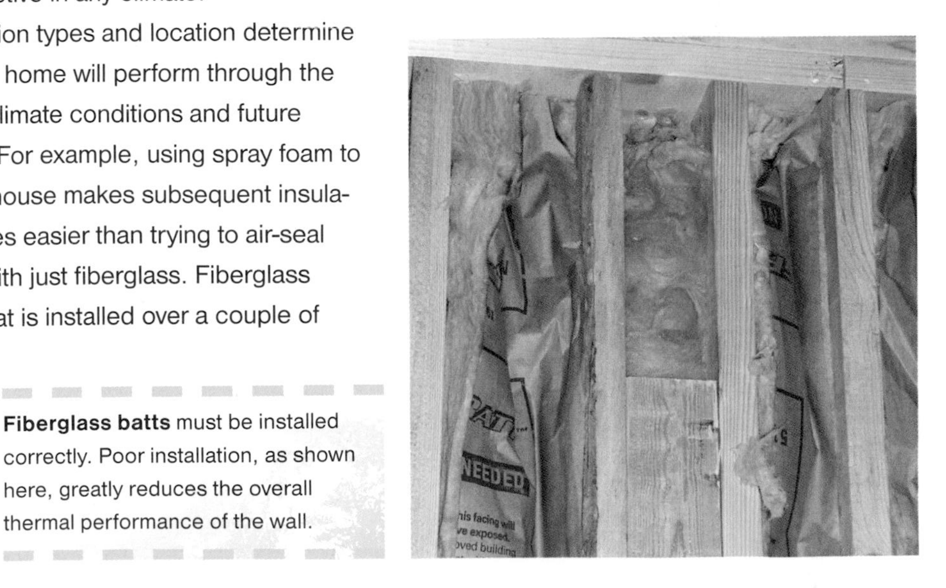

Fiberglass batts must be installed correctly. Poor installation, as shown here, greatly reduces the overall thermal performance of the wall.

GETTING TO ZERO

Insulation is the best investment you can make today. Compared to natural gas prices doubling in the last decade, insulation is relatively inexpensive, and it is a fixed construction cost rolled into a mortgage. If insulation costs amortized over the term of a second mortgage save $1 more in utility bills than the monthly payment on the loan, there is a positive cash flow from the first month. Energy prices will continue to rise in the years to come. The rewards of investing in superinsulated houses will be well worth it in the next decade.

ing, or other obstacles already installed. Fiberglass requires conscientious air sealing before installation: All vertical and horizontal penetrations from the envelope need to be foamed to prevent air movement. The attic/second-floor connection requires perfect air sealing because fiberglass itself does very little to resist air movement. Proper installation requires cutting the batts around every wire, electrical box, and pipe in the wall. Shoving the batt around an obstacle ruins the integrity of the wall and also creates the potential for mold.

Fiberglass insulation comes in different densities. A 2×4 wall can have an R-11, R-13, or R-15 value, depending on the density of the material. Higher-density fiberglass is more useful because it has a greater R-value per inch, which can save money when creating deeper cavities with framing.

Cotton batts

Cotton batts are made from recycled materials. To work effectively, they require perfect installation around all obstacles in a wall cavity—just like fiberglass batts. A downside is that the material is very difficult to cut. While fiberglass can be cut using a utility knife, cotton just shreds when using a knife. Two tricks for the perfect cut: An electric knife slides right through the batt, and a manual hedge trimmer is great for straight cuts.

Cellulose

Cellulose insulation is gaining ground in conventional construction. It has high recycled content and is more affordable than foam (discussed next). Make sure to use only borate-treated cellulose, as any moisture that gets into ammonium sulfate-treated cellulose will cause it to off-gas ammonia. Also, it takes a professional to install cellulose in walls. For ceilings, loose fill helps reduce air flow through the insulation, although it is still permeable.

Cellulose is installed dry as loose fill in walls and is kept in place behind netting. Here again, perfection is required. If it is too dense, it fills out the netting and makes drywall difficult, if not impossible, to install. If it is too loose, it will settle and leave gaps at the top of the cavity. When installed at the correct density, it is an effective insulator and has some great acoustical properties.

Cellulose can also be sprayed on, mixed with an acrylic binder that holds it firmly in the wall cavity. It can be sprayed in as deep a wall cavity as necessary and stays stable. It has the advantage that it fills around wiring, plumbing, and other obstacles in the wall cavity so it is a better air seal yet is still permeable. Like its dry counterpart, it also has some great acoustic properties.

Spray polyurethane foam

Spray foam tends to be the insulation of choice for zero energy home builders, primarily because of its superior air-sealing properties. Like spray cellulose, spray foam also fills around pipes and wires in the walls for a perfect seal. There are two forms of spray polyurethane foam: closed-cell and open-cell.

Closed-cell foam

Closed-cell foam has a higher R-value and is more water resistant than open-cell foam. Sometimes called 2 lb. foam because it weighs 2 lb. per cubic foot, closed-cell foam is actually a commercial roofing product with a permeability of less than 1. It has structural properties that increase the strength of a wall by 30 percent and improves the connection between walls and roofs four times over toe-nailed trusses. It is a great product for high-wind

Cellulose is made from recycled newspaper; it is fire-retardant.

Foam insulation is blown in place and expands to fill the full cavity. The excess must be removed so that drywall can be installed.

load areas or for tornado- or hurricane-prone zones. Over time, the blowing agent that expands the foam (hydrofluorocarbon) off-gasses. This process decreases the R-value until it stabilizes. Closed-cell foam has a stable R-value of 6 per inch, although some manufacturers claim R-6.5 to R-7 (the R-value when it is fresh).

Closed-cell foam should be sprayed only 2 in. to 3 in. per pass rather than all at once because of the heat it creates as it sets. Cost-conscious builders use one pass of closed-cell foam for air sealing and then fill the rest of the cavity with cellulose or fiberglass.

Open-cell foam

More widely used than closed-cell foam, open-cell foam has a density of 1/2 lb. per cubic foot. It expands rapidly—up to 100 times its initial volume when applied—but as it is expanded with water, there is no off-gassing. It is vapor permeable, which means it dries faster, but unlike closed-cell foam it is not water resistant. As a result, its R-value is only 3.5 per inch (about the same as fiberglass or cellulose). Its advantage over fiberglass or cellulose is that it still performs as a great air barrier. Open-cell foam is priced considerably lower than closed cell and can be sprayed to fill the entire wall cavity.

Testing the Building

Once all the insulation has been installed and air sealing is completed, the building needs to be tested to make sure that it performs as designed. This step typically identifies areas where insulation has been missed, air sealing is insufficient, or mechanical systems are

R-VALUES AND AIR INFILTRATION

An R-value is a measure of a how much time it takes 1 Btu of energy to pass through a given material. While a lot of time is spent talking about R-values, it's also important to consider insulation's ability to prevent air and moisture infiltration.

Although fiberglass insulation is still the most common product used, spray foam or cellulose insulation are much more effective products for a variety of reasons. For fiberglass to perform well, it must be installed correctly. If voids and gaps are left unfilled, heat and sound will travel, negating much of the insulation's protective benefits. For instance, fiberglass is generally assigned an R-value of approximately 3.5 per inch, but it will achieve that R-value only if tested in an absolute zero wind and zero moisture environment.

When evaluating an insulation product, it is important to look at more than just the R-value of the product. Because the available R-value ratings for insulation materials do not account for the affects of air movement, an insulation system—such as polyurethane foam—provides much higher performance than its actual R-value rating when properly installed. In comparing the cost of closed-cell foam against fiberglass or other products, keep in mind that the lack of air migration in a closed-cell system better maintains the integrity and performance of the thermal envelope and therefore yields a far better R-value.

Spray foam in clerestory rafters makes the passive solar aperture as high performing as the rest of the roof.

not functioning up to specs. A zero energy house is just that: It scores 0 or below on the home energy rating system (HERS) scale. It is difficult to get a 0 score, but most near zero energy homes get below HERS 20, which is excellent, given that a conventional home built to code has a HERS score of 100. (For more on the HERS scale, see "What is a HERS Rating?" on p. 100.)

Testing for remodels and new construction is similar, with one exception. For a zero energy retrofit of an existing building, the place to start is with a computerized analysis of the utility bills, which can help separate out the energy used for heating, cooling, water heating, and miscellaneous electrical loads. Then, the testing procedure includes the tests and checks discussed in the following sections. It is important to have the house tested before construction starts and then again when the home is finished.

Blower-door test

A blower-door test puts the house under negative pressure to determine its relative airtightness and to locate specific points of air leakage. The blower door is a fan that exhausts air to a specific pressure inside, typically 50 pascals, the equivalent of 10 mph to 15 mph wind speed. The fan is positioned in an expandable frame that is sealed into an open door jamb. Used in conjunction with a smoke stick test, the blower-door test allows you to actually see where air is leaking. The test is best performed before the drywall is hung so remediation can take place where the air leakage is discovered. This is a critical step for zero energy homes. Conventional homes with high-powered HVAC equipment can get away with some air leakage. When a home is primarily heated with solar and equipment waste heat, airtightness becomes crucial.

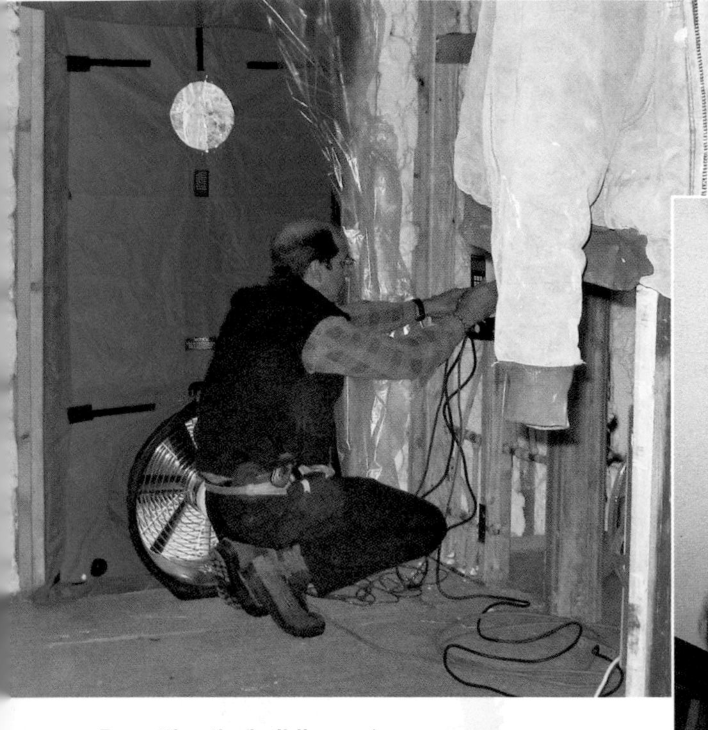

A smoke test is useful for finding air leakage around light fixtures and windows.

By putting the building under pressure, a blower-door test determines how much air a home is leaking.

Insulation check

An energy audit is another important step in evaluating a home's weak spots. An audit covers many things, including inspection of the attic, basement or crawl space, and walls to evaluate the quality of the insulation installation. In conjunction with the blower-door test and an infrared camera test (see p. 41), this step ensures that the thermal barrier is intact.

Heating, cooling, and hot-water heating appliance assessment

The water heater, furnace or boiler, and air-conditioner should all be inspected for efficiency. Testing the ducts with a duct blaster, if applicable, determines the integrity of the duct connections. Many auditors are also able to perform a worst-case combustion air zone test to determine if appliances are venting properly (such as when all of the exhaust fans in the home are operating at the same time) to prevent back-drafting of carbon monoxide down flues. This test also identifies whether the HRV or ERV is functioning properly.

Duct blaster test

A duct blaster test uses a device that pressurizes the ductwork and determines air leakage through pressure loss. If you are using ductwork for conditioned air, it needs to be airtight.

EnergyStar requires less than 6 percent air leakage, which is a tough but obtainable standard. All joints in ductwork must be sealed with mastic to eliminate air leakage, and panned ducts should never be considered. (A panned duct is one in which a piece of sheet metal is screwed to the bottom of a joist to reduce sheet metal work.) Panned ducts trap moisture in the joist space and are always leaky. When you are dealing with subtle air flows in a zero energy home, it is critical to have air movement only where and when you want it. Air losses with heat-recovery ventilation can reduce the efficiency of the equipment.

Windows and doors

An audit should also include an inspection of the windows and doors in a home, both for airtightness and energy efficiency. On occasion, the expanding foam between the rough opening and the window can bind the frame so that it doesn't open and close appropriately. Doors may have to be adjusted at the hinges to make sure they seal properly. This is where the smoke stick test comes in handy.

Lighting and large appliance evaluation

An energy auditor will assess the efficiency of the lighting system, refrigerator, washing machine, freezer, and other appliances. Each appliance should be plugged into a watt meter to determine and record how much power is required to run it. The results will affect the size of the photovoltaic array required to meet the total load (see chapter 3).

Infrared camera scan

Infrared cameras (IRCs) allow you to look through walls and ceilings to check for missing insulation and locate air leaks. The IRC scan is best conducted when there is a temperature difference between inside and outside of at least 20°F. The greater the temperature differential, the more dramatic the photos.

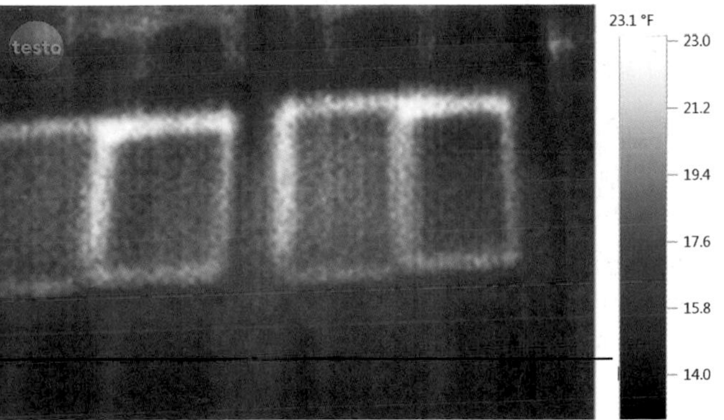

An infrared camera test shows heat loss around windows and also detects places where there is missing insulation.

Putting It All Together

Designing the envelope as a system is required rather than suggested. It takes careful design, computer simulations, trial and error, and rethinking to get it right for your climate. For each of the cases study houses shown in this book, the designers and builders have taken different approaches to tightening up the building envelope. In general, the simpler the approach and the fewer the cavities yield the best results.

Case Study

High Heat

Florida research house built to ease peak loads on electric utility

On June 18, 1998, as afternoon temperatures climbed to 100°F in central Florida, Lakeland Electric recorded its maximum peak load for the summer, a record 1-minute period in which demand hit 578 megawatts. As oppressive as it was, the afternoon was just about perfect for comparing two nearly identical houses in the Windwood Hills development.

SPECIFICATIONS

House size: 2,425 sq. ft.

Average cooling degree days at site: 3,500

Wall and roof construction: Concrete block walls, truss roof

Insulation type: 1¼ in. polyisocyanurate (R-10) on wall exterior; R-30 blown fiberglass in ceiling

Windows: Double-pane, low-e, argon filled with thermally broken vinyl frames (Solar Heat Gain Coefficient of 0.38, U-valve 0.35)

Photovoltaic capacity: 4kW in two arrays

Heating source: Propane

Air-conditioning: 2-ton (SEER 14.4) variable speed

Cost per square foot: Total building costs not available; energy upgrades $14.51 per sq. ft. not including utility-owned photovoltaics

Designer: Florida Solar Energy Center

General contractor: Strawbridge Construction

At the time this research house in Lakeland, Florida, was built in 1998, the net zero energy boom was still years in the future. This 2,452-sq.-ft. house looks much like its neighbors, but its energy performance is a whole different story.

Both were about 2,500-sq.-ft., one-story houses built by the same company within the previous year. One was of standard design and materials for the area, the other was a research house equipped with rooftop photovoltaic (PV) panels and a variety of features aimed at reducing energy use. Inside the conventional house, the 4-ton air-conditioner labored from 11:00 in the morning onward trying, unsuccessfully, to keep the indoor temperature at the 76°F set point. Temperatures rose inside to a hair under 80°F, despite a 4.5kW power draw by the air-conditioner.

Just down the street, the temperature inside the air-conditioned research house was 74°F. At the time of peak demand, the net electrical draw was 225W, about what two incandescent light bulbs would use and 95 percent less than the standard home.

To be sure, its energy performance is something less than net zero. But more than a dozen years ago, when net zero building was still a gleam in some building scientist's eye, this house was a huge step forward. Its designers at the Florida Solar Energy Center helped point the way for the many researchers who would follow.

Beating Standard Construction Performance

The Lakeland house was never explicitly designed as a zero energy building. More precisely, designers were testing ways of building houses that, with the help of photovoltaic systems, would create very small loads on the utility grid during the summer when demand is greatest.

Danny Parker, principal research scientist at the Florida Solar Energy Center, says the design was based on computer simulations the center had run in the early 1990s. For money to test the ideas, the center turned to the Florida Energy Office and Sandia National Laboratories, which paid for the added energy features in the research house. The same construction company built both houses.

Reduced electrical demand holds huge promise for utilities. According to a Solar Energy Center report on the project, Florida at the time was seeing roughly 100,000 new houses a year, and a typical three-bedroom house in the central part of the state would use about 15,000kWh of electricity annually. Between 4:00 P.M. and 6:00 P.M. on an average summer day, peak demand at each of these houses would be 4kW. Customers living in high-efficiency homes would reap direct financial benefits with lower bills, while utilities could hope they would not be forced to build new power plants.

But to get there, standard building practices clearly would have to change. Wall insulation in houses built from 8-in. concrete block, which was typical, usually amounted to an interior radiant barrier (Fi-Foil® reflective insulation), the equivalent of about R-4. Wraparound porches that had traditionally shaded windows and exterior walls from direct sunlight in Florida Cracker homes had given way to skimpy roof overhangs as builders looked for ways to cut costs. Air-conditioning ducts insulated to R-6 were left in unconditioned attics where summer temperatures could exceed 130°F.

MORE THAN A REGIONAL RESEARCH CENTER

Three decades after it was created, the Florida Solar Energy Center has emerged as an important center for national as well as regional research on energy conservation and solar energy systems.

The center conducts research in a variety of building science fields, including windows, heating and air-conditioning equipment, insulation, appliances, and air distribution systems. In all, it has a staff of about 150 and a $3 million annual operating budget from the University of Central Florida, along with grants and contracts worth up to an additional $12 million a year.

For zero energy builders, the center's evaluations of solar hot-water and photovoltaic systems are especially valuable. Its website (www.fsec.ucf.edu) includes a listing of solar water-heating equipment that has been tested and certified, along with excellent background information on the different types of systems available and pointers on how to size a domestic hot-water system.

LOTS OF LITTLE STEPS MEAN BIG ENERGY PERFORMANCE

A variety of energy-conserving measures at the Florida research house sharply reduce electricity use when compared with a nearly identical comparison house in the same neighborhood. When air-conditioning at the control house was drawing more than 4kW in the later afternoon on two of the days represented in this chart, the research house was using roughly one eighth as much. Prechilling the house before peak electrical loads also helps reduce the strain on the utility's grid later in the day.

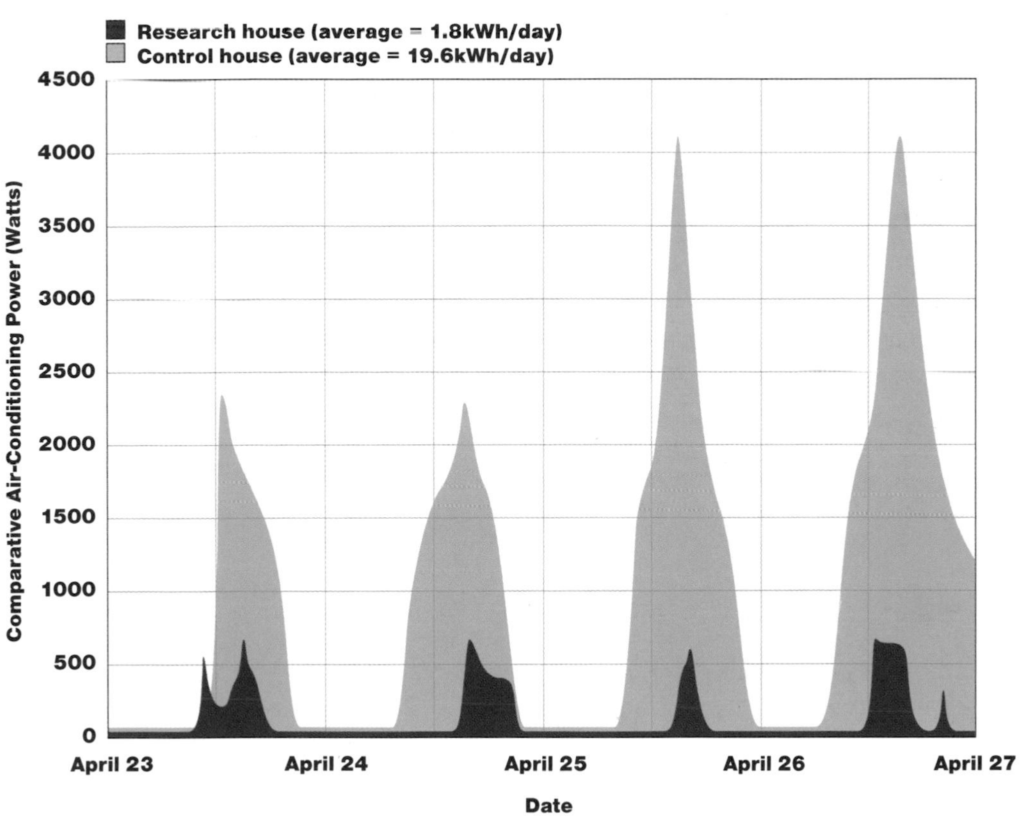

Adapted from Florida Solar Energy Center

These were the business-as-usual features that went into the conventional comparison house in Woodland Hills. At the research house, nonstandard features included a 4kW PV system, reflective white roof tiles, 3-ft.-wide roof overhangs, high-performance windows, compact fluorescent recessed lighting, high-efficiency air-conditioning, and a solar hot-water collector. Insulation in the exterior walls was increased, and air ducts were brought inside the conditioned space.

Upgrades to the building shell seem modest, especially by northern standards, but collectively they had a huge effect on energy consumption. At the research house, insulation was moved to the outside of the block. Instead of a reflective barrier, builders wrapped the exterior in a 1 1/4-in.-thick layer of rigid polyisocyanurate foam (R-10). Changes in traditional roofing had an especially big impact.

Tighter construction resulted in less air infiltration. A blower-door test measured leakage at 4.9 ACH at 50 pascals of pressure, compared with 6.3 ACH under the same conditions at the comparison house. NACHs were estimated at 0.16 in the zero energy test house and 0.21 in the conventionally built house.

Photovoltaic panels with a peak capacity of 4kW are not enough to offset consumption entirely, but they help ease the load on the local electricity utility at times of highest demand.

In addition to photovoltaic panels mounted in two separate arrays, the house also has a solar hot-water collector. A propane-fired water heater provides the backup.

KEEPING IT COOL

Using white concrete roof tiles instead of standard gray-brown asphalt shingles was especially effective in keeping attic temperatures down. Inside the attic of the conventional house on that June 18 utility peak day, temperatures reached nearly 138°F, while in the research house the temperature was about 100°F, the same as the temperature outside. Both houses had R-30 of blown-in fiberglass in the ceiling.

Because ducts carrying 60°F conditioned air in the conventional house went through the attic, Parker estimated that a half ton of cooling capacity was lost before the cool air reached its destination. Keeping ducts inside the conditioned space in the research house solved that problem and helped designers reduce the capacity of the air-conditioning system from 4 tons to 2 tons (48,000 Btu to 24,000 Btu).

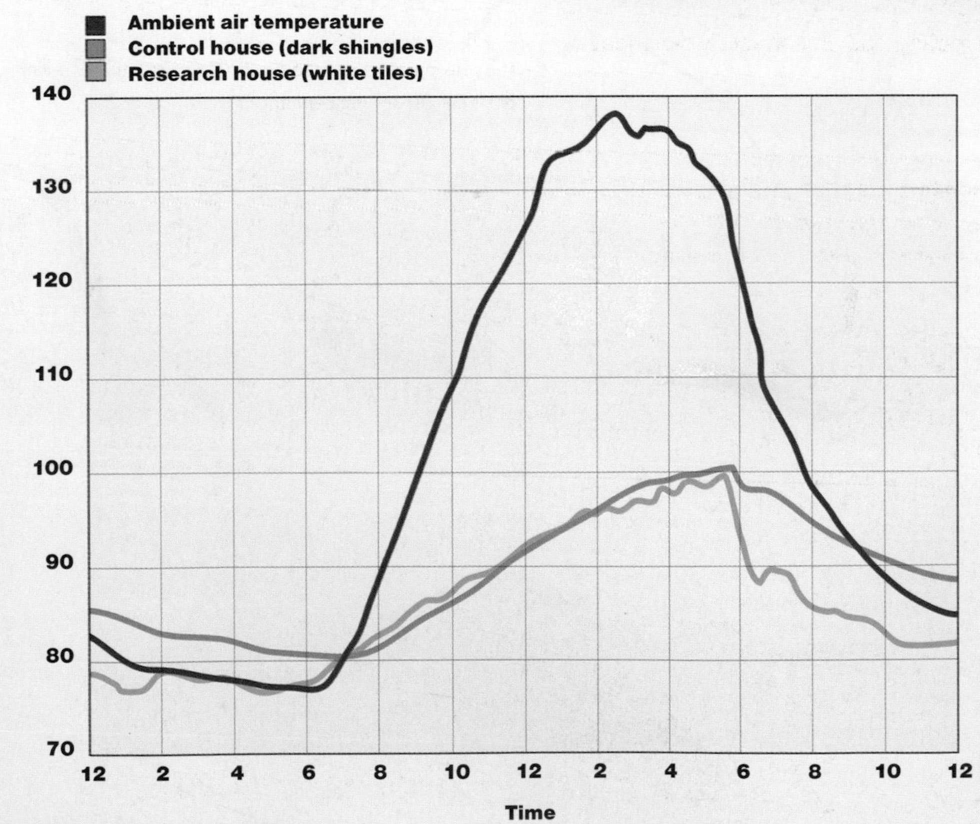

Adapted from Florida Solar Energy Center

Splitting the PV system for maximum gain

Instead of placing all of the photovoltaic modules on a south-facing roof, designers split the array in two: 36 of the 75w Siemens modules went on the south-facing roof and the remaining 24 modules were on a west-facing roof. The idea was that the south-facing array would perform well in the middle of the day, and the west-facing array would be more efficient later in the day when electrical loads reached their peak and the utility would gain the most from a reduced load. Nominal output was 1.3kW on the west-facing array and 2.7kW on the south-facing array (actual measured performance was less).

Over a 5-month stretch from April to August, total electrical loads in the house averaged 22kWh per day, while the PV system delivered 15.6kWh per day to the grid—about 71 percent of the total consumption. During the day, there was essentially no draw on the grid.

To reduce electrical loads, the house was equipped with a propane stove and clothes dryer and a propane-fired backup to the solar hot-water collector. These appliances use 54 gal. of propane per year, the equivalent

Rather than group all solar panels in one array, designers split them into two groups. A west-facing array gains in efficiency during the late afternoon, just as the local electric utility is beginning to see peak electric demand.

of 1,511kWh of electricity. Given a net electrical use of about 5kWh a day along with the propane for appliances, the house would have some distance to go to reach net zero performance (although at the national average today, net electrical consumption would cost about 80 cents per day).

The Lakeland research house verified predictions by researchers at the Florida Solar Energy Center and, says Parker, became a forerunner for the U.S. Department of Energy's zero energy program. And with a few tweaks to the design—more floor tile and less carpet to allow the ground to be a more effective heat sink, sealed ceiling light fixtures, and a little more PV capacity—zero energy performance would have been tantalizingly close.

An oddity at the time the house was built, the inverter and grid connections for the photovoltaic panels are mounted on the east wall of the house.

Some of the solar features are passive, including the light-colored roofing tiles. They helped reduce attic temperatures substantially when compared to a neighboring house with a dark asphalt roof. Lower attic temperatures mean a lower cooling demand.

Case Study

A Butterfly Home in the City

An upside-down roof hides solar panels, boosts efficiency

The upturned edges of the roof on Michael Yannell's Chicago net zero home certainly help set it apart from its urban neighbors. But what its architects call a butterfly roof is more than an architectural flourish. It's one of several features designed to improve indoor comfort and lighting, and minimize energy consumption.

SPECIFICATIONS

House size: 2,675 sq. ft.

Average heating degree days at site: 6,493

Wall and roof construction: Concrete block and 2×6 wood-framed walls; 2×10 roof joists

Insulation type: Rigid extruded polystyrene, spray-in polyurethane foam

R-values: South, east, and west, R-33; north, R-35; roof, R-57

Windows: Triple-glazed, argon-filled, low-e; U-0.24

Photovoltaic capacity: 9.8kW

Heating source: 5-ton ground-source and air-souce heat pump

Air-conditioning: Heat pump

Cost per square foot: Not available

Designer: Farr Associates, Jonathan Boyer design principal

Energy Modeling: dbHMS

General contractor: Goldberg General Contracting, Inc.

Two separate buildings linked by an entry and foyer enclose a common courtyard. One wing contains public spaces, the other the bedrooms and bathroom.

Final working plans for the 2,675-sq.-ft. house, designed to meet Leadership in Energy and Environmental Design (LEED) Platinum standards, followed nearly 9 months of computer energy modeling that reflected the architects' efforts to marry performance and design. Because energy systems weren't completely up and running until July 2009, it will be midway through 2010 before designers know whether the house is a true net zero performer. But with a robust solar hot-water system and 10kW photovoltaic array, designers built in a comfortable energy margin.

In addition to its energy systems, the house also incorporates several green features, such as a gray water collection system, sustainably harvested materials, storm water management, and turf-free xeriscaping on the lot to minimize water consumption.

Roof and Floor: Form Follows Function

The home is designed as two wings connected by an entry/foyer that face a central courtyard. The south wing houses the kitchen, dining room, and living room in an open-plan design; the north wing includes a bathroom and bedrooms.

GREEN FEATURES FOR LEED PLATINUM

Architects Farr Associates designed this Chicago house to meet LEED Platinum standards, the highest designation under the national green-building program administered by the U.S. Green Building Council. Here are some of the house's green features:

- **A basement cistern where rainwater from the roof can be stored and later used to irrigate outdoor plants.**
- **Permeable hardscape materials.**
- **Sustainably harvested materials. Nearly all of the structural wood used in the house is certified under the Forest Stewardship Council program. Cabinetry is made from sustainably harvested hardwoods, and kitchen counters include surfaces made from recycled newsprint.**
- **Outside, the turf-free landscaping includes native and noninvasive plants that meet National Wildlife Federation requirements for a Certified Backyard Habitat.**
- **EnergyStar lighting and appliances, low-flow plumbing fixtures, and wall paint free of volatile organic compounds (VOCs).**
- **More than 90 percent of construction waste was recycled.**

The floor plan effectively doubles the south-facing area of the house and with it the amount of solar energy that can be harvested in the winter. A wall of windows on the south face of each wing also provides ample natural light inside the house.

One of the advantages of the butterfly roof is that photovoltaic and solar hot-water panels are mostly hidden from view. Insulated to R-57, the roof also is at an ideal angle for PV efficiency, while its inverted shape collects water that can be used for landscape irrigation with the help of a 550-gal. cistern in the basement. The roof is insulated with a combination of extruded polystyrene and closed-cell polyurethane spray insulation.

Inverted roofs help disguise solar hot-water and photovoltaic panels from the street while gathering rainwater that can be stored in a cistern and used for landscape irrigation. Deep roof overhangs help the interior stay cool in summer.

Evacuated-tube solar hot-water collector are designed to provide all the hot water th household needs and give a boost to the radiant-floor heating system as well.

On the north side of each building, walls are made from 8-in. textured concrete block to the inside and an insulated 2×6 stud wall to the outside. In winter, the massive walls gather heat from the sun through the banks of south-facing windows during the day and release it gradually at night. Between the block and the stud wall is a 6-in. air space that is used as an air plenum. Heated or cooled air passes over the mass walls and is recirculated through the house to keep it comfortable. Stud walls to the outside are insulated with closed-cell polyurethane foam.

The foundation is surrounded by 2 in. of extruded polystyrene insulation with an additional 2 in. below the basement slab.

Ventilation Starts with Many Windows

There are more than 100 triple-glazed windows in all, including a band of clerestory windows that help keep the interior bright. The majority of the windows are operable to make the most of natural ventilation. Few windows, however, are located on either east or west walls to reduce the amount of solar gain during Chicago's hot, sticky summers.

This unique wall construction incorporates a 6-in. plenum for air circulation between the 2×6 framed wall on the outside and a block wall to the interior.

South-facing walls of triple-glazed windows let sunlight warm exposed masonry walls and a concrete floor. During the summer, broad roof overhangs block the sun and help prevent overheating. Most of the 100+ windows are operable.

Windows are located low on the cool north side of the house for air intake. Stale indoor air is pushed out of the house with roof ventilators. During the winter, dampers are closed and vents are covered. When natural ventilation or mechanical ventilation isn't enough to keep the house cool, an air-conditioning system kicks in.

Grid-Tied Solar Electric System Designed to Overproduce

The home is designed to run completely on renewable energy sources. Energy modeling predicted total annual electricity use at 12,689kWh and production from the 10kW photovoltaic system at 18,000kWh per year, giving the house a comfortable margin of production. Monitoring started on July 1, 2009.

The HVAC system is a 5-ton (60,000-Btu) system that incorporates both ground-source and air-source heat pumps. The water side, connected to three 260-ft.-deep wells, heats a radiant-floor system while the air-source heat pump is useful for bringing the house up to temperature quickly.

The home also has an evacuated-tube solar hot-water system, which is designed to provide all the domestic hot water as well as to preheat water for the radiant-floor system. An energy-recovery ventilation system provides fresh air.

During the summer, stale inside air is exhausted through turbine ventilators in the roof. The house also is ventilated with an energy-recovery ventilator.

SALVAGING GRAY WATER FOR REUSE

Global freshwater resources are threatened by rising demand, making the use of recycled water instead of potable water an attractive option when possible. On average, between 50 percent and 80 percent of the water used in a house goes to showers, sinks, and the laundry. A gray water system is a method of reusing this water before it enters the sewer system. After its initial use, water is filtered and stored. Later, it can be used either to flush toilets or for irrigation. Gray water can come from a variety of sources, but it can't include toilet waste or water from kitchen sinks or dishwashers.

There are several systems available, including AQUS®, which is suitable for remodels; ReWater® system, for irrigation only; and the Brac recycling system, which is best suited for new construction. A newer product on the market, Water Legacy, disinfects water with a combination of UV and hydrogen peroxide.

Gray water systems are not permitted in all areas. Check with your local building department before investing in this technology.

Mechanical systems include a geothermal heat pump and a gray water storage and filtration system, the city of Chicago's first residential project to include such systems.

Letting the Materials Speak for Themselves

The massive 8-in. textured concrete block walls are exposed to the inside. Instead of using traditional soffits along roof eaves, the designer finished roof overhangs with tongue-and-groove wood decking, a material that could safely be exposed to the weather. Glulam structural beams on the house interior were left exposed, in keeping with the designer's intent to leave structural elements of the building exposed where possible. When available, materials were locally sourced and made from recycled content.

Yannell's goal was to build a net zero home using off-the-shelf materials, not products that were exclusive or difficult to acquire. Although the house incorporates detailing that's far from standard, both the designer and the homeowner liked the "simple is better" approach. When presented with options, the designer tried to pick the most straightforward solution that provided Yannell with a home that was beautiful as well as easy to operate.

The clean crisp interior of this home uses many recycled and environmentally friendly products.

Case Study

Outside-In Retrofit

Adding insulation and new windows to the shell while leaving the house interior untouched

Retrofitting an existing house for substantially better energy performance has its challenges, especially when the building is nearly a century old and home to both its owners and four student tenants. If the house is completely gutted from the inside, the project begins to resemble new construction in both cost and complexity. Staying in the house during construction can get tricky, and period trim and detailing could be difficult or impossible to preserve.

SPECIFICATIONS

House size: 2,966 sq. ft. (conditioned space)

Average heating degree days at site: 5,600

Wall and roof construction: Existing 2×4 frame

Insulation type: Spray foam added to existing cellulose

R-values: Walls, R-39 in most areas; attic, R-60; foundation walls, R-25

Windows: Double-pane, low-e added over existing single-pane

Photovoltaic capacity: 5.25kW

Heating source: High-efficiency condensing gas boiler

Air-conditioning: None

Cost per square foot: $47 (retrofit costs calculated for conditioned space)

Designer: Steve Baczek, Architect

HERS Rater: Mike Duclos, Energy Efficiency Associates

General contractor: Byggmeister Design Build

Older homes pose special problems during energy-oriented retrofits, but given the number of existing homes in the United States, solving these challenges is crucial.

Cador Pricejones knew all that as he pondered energy upgrades for his 1914 home in Somerville, Massachusetts. Pricejones, the production manager at Byggmeister, a Boston-area remodeling and design company, had a better plan: Tackle the job from the outside and leave the house interior unscathed.

By adding a thick layer of insulation on the outside of the house, all of the wood framing would stay warm during the winter, reducing the risk of condensation inside the walls. Thermal bridging, the loss of heat through the wood frame, would be much less of a concern. Although net zero performance wasn't a realistic goal, a more comfortable house with much lower energy consumption was definitely in the cards.

An Energy Hog Worth Keeping

Pricejones was starting with a house with an unofficial HERS rating of 119, meaning the house used almost 20 percent more energy than a house built to current energy codes (see "What Is a HERS Rating?" on p. 100). Most of the windows, siding, trim, and interior were original, and the family liked the house well enough to want to stay put rather than sell and start again.

Energy use for the 2,966 sq. ft. of living space in the house was about 88,000 Btu per square foot per year, slightly higher than the average for two- to four-family homes in the Northeast United States. A calculator for the Thousand Home Challenge (see below) recommended energy consumption of 18,000 Btu per square foot per year, a reduction of about 80 percent.

Pricejones had done the easy things: installing energy-efficient appliances and lighting fixtures, tracking down phantom electrical loads, and insulating the attic. Still, the house needed more drastic intervention, and that was going to be complicated. Four graduate students lived on the first floor and part of the second; Pricejones's family took the other half of the second floor and the space at the top of the house. Those circumstances made an even stronger case for working from the outside.

Adding Insulation to the Outside

Existing walls were framed with 2×4s and insulated with blown-in cellulose. The building's 36 windows were all single pane, typical for a house of its age. Pricejones decided not to disturb either the siding or the existing windows. Instead, he attached new double-hung windows to the casing of the existing windows and then created new exterior framing by screwing 2×3 studs to the outside of the building through 3-in.-thick blocks of rigid foam insulation.

The studs created a new wall cavity on the exterior of the building, which Pricejones then filled with 4 in. of closed-cell polyurethane foam. A small air space between the face of the new insulation and the face of the studs made a shallow air space behind the siding, allowing the back of the siding and trim to dry should it get wet. When added to the cellulose insulation already in the existing walls, Pricejones got a total wall R-value of nearly 40.

THOUSAND HOME CHALLENGE

Finding workable approaches to reducing energy consumption in existing houses is the goal of the Thousand Home Challenge. The fledgling effort grew out of a conference on carbon neutrality convened by Affordable Comfort, Inc. (ACI), a Pennsylvania-based nonprofit, in 2007.

According to Linda Wigington, director of special projects for ACI, one of the conference participants suggested 1,000 homes was a big enough number to demonstrate that steep energy reductions are possible in more than a one-off, custom approach.

Efforts are directed at reducing total annual site energy in North American homes by between 75 percent and 90 percent while addressing a variety of other issues, including indoor air quality and durability. The program seeks to develop performance standards that are "easy to understand and use, and accommodate a wide range of climates, housing types, and fuels."

By summer 2009, the program was still in a pilot stage. With its eventual launch, ACI hopes to stimulate collaboration, develop a specialized workforce, develop energy retrofit approaches that can be replicated for various common house styles, and help guide lifestyle choices that affect energy use.

ACI says more than 50 groups are collaborating on the effort, including the Canada Mortgage Housing Corporation, the Lawrence Berkeley National Laboratory, the National Renewable Energy Laboratory, Oak Ridge National Laboratory, and the Passive House Institute U.S.

One of the first steps in a deep-energy retrofit at this Massachusetts two-family home was to attach new windows to the outside of the house. They were installed directly over the existing casing. The single-pane window already in place became a sort of interior storm window.

Without removing the old siding, Pricejones screwed 2×3 studs to the outside of the house through blocks of rigid foam insulation. This approach creates new wall cavities for high-density polyurethane foam and virtually eliminates thermal bridging through exterior walls.

In the end, the house has two sets of windows, the original single-pane units and new double-pane units installed directly over them. The arrangement isn't as odd as it may sound. Pricejones says it's not unlike having storm windows, and the combination of the existing windows with new double-paned low-e units means a much tighter building envelope.

With 4 in. of closed-cell foam sprayed over walls that had already been insulated with cellulose, total R-values are about 39 in most places. Equally as important, the foam is a waterproof and effective air seal.

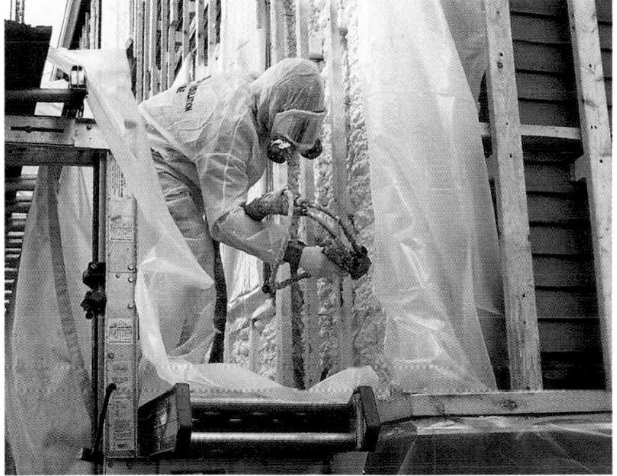

At the sill, Pricejones removed the first 4 in. of siding and attached a horizontal nailer to provide support for trim and a solid attachment point for the bottom of the new studs.

In the attic, Pricejones sealed all plumbing and wiring penetrations and added more cellulose to bring the total R-value to 60. In the basement, he framed a wall with 2×4 studs on the flat, leaving a gap between the framing and the foundation wall, and sprayed in 4 in. of closed-cell foam against the foundation wall. Studs provided the backing for moisture-resistant drywall. The basement slab remained uninsulated, in part because it would have been a logistical hassle to move everything and also because of the limited value of insulating a below-grade area that is unconditioned.

Better air sealing and a new cocoon of polyurethane insulation made whole-house ventilation a necessity. Pricejones decided to go with two systems, one for the upper part of the house where he lived with his family and another for the tenants. The heat-recovery ventilators are rated at 70 cu. ft. per minute.

Upgrading the Mechanical Systems

Pricejones replaced two older boilers and two water heaters in the building with a single high-efficiency condensing gas boiler with an indirect hot-water storage tank. The sealed-combustion boiler does not require a chimney. There is no central air-conditioning system.

Renewable energy systems, which Pricejones is planning to add, will include three flat-plate solar hot-water collectors designed to produce 100 percent of the domestic hot-water needs during the summer. During the winter, when the amount of sunlight falls

Although the house wouldn't look dramatically different after the retrofit, one giveaway, if you knew where to look, was the deep overhang at the top of the foundation wall.

SAME LOOK, MUCH BETTER PERFORMANCE

Working with an existing building inevitably means making compromises, and Cador Pricejones found there was no exception as he worked on his own home.

There wasn't enough roof overhang on some parts of the house to add a new stud wall and spray-foam insulation. In those areas, Pricejones had to settle for a little less thermal performance. Adding insulation to the basement slab was just too much of a hassle for the limited benefit he would have seen. To make the addition of rooftop solar panels practical, a beautiful slate roof had to go.

But in the end, the benefits of the energy retrofit far outweighed whatever challenges he faced along the way. New windows and spray-foam insulation sealed the air and thermal leaks common to old houses and made it much more comfortable. Energy use is way down. And even with all of the changes, the house essentially looks just the way it did when Pricejones started. So good, in fact, that he was planning on applying for a Somerville historic preservation award.

dramatically, the system will produce a portion of the hot water.

Also on order was a 5.25kW photovoltaic array for the roof. Thanks to low electrical consumption, Pricejones expects that the PV system will have enough capacity to meet all of his family's needs plus some of the tenants'.

Pricejones expects the house's HERS rating to fall from 119 to roughly 52, not including any of the renewable energy contributions. Once the solar hot-water and PV systems are factored in, he hopes to see the HERS rating fall to somewhere near 40 (a net zero house would have a HERS rating of 0).

GETTING TO ZERO

One of the keys to reducing atmospheric carbon dioxide and averting catastrophic climate change is to cut energy consumption dramatically in existing houses. According to owner Cador Pricejones, "There are 120 million homes in America and we need to figure out how to retrofit them for the new reality that we live in. We cannot just choose to move to the country and build a net zero home and let someone else deal with existing houses."

With new wall cavities filled with foam, the house is ready for siding. Because the foam forms an impermeable layer, no housewrap or other barrier was needed for an effective weather shield.

New siding completes the project on the outside. A shallow space between the back of the siding and the foam insulation allows siding and trim to dry to the inside if they become wet.

2 Passive Solar Design

Sound familiar? The Greeks faced an environmental crisis in the 5th century B.C.E. Wood was in high demand for ships and buildings, making the primary heat source for comfort—charcoal—increasingly scarce and expensive. Most of the Greek peninsula and the islands had been denuded so they were forced to import wood from faraway places. This drove the ancient architects to reevaluate how they built their buildings.

According to Socrates, "In houses that look toward the south, the sun penetrates the portico in winter, while in summer the path of the sun is right over our heads and above the roof so that there is shade." Thus was born the idea of passive solar heating and summer shading.

As we look back across American architectural history, we see that early settlers also wisely used the sun for comfort. Early Native American settlers of the American Southwest were very conscious of the movement of the sun and built their pueblos so that the community had access to the southern sun. The people of the Anasazi culture that populated the Four Corners area of Colorado, Utah, New Mexico, and Arizona lived in large open caves that faced south. They built their dwellings recessed far enough from the opening that the tall cave roof shaded the homes in the hot summer months.

To this day, entire communities take advantage of solar heating and summer shading. Shown here are houses on the ancient Greek island of Santorini.

All over what is now the southwestern United States, Native American settlements and pueblos were built using massive adobe materials to absorb heat in winter and protect occupants from hot summer sun.

A more contemporary example is the saltbox design favored by the early colonists in New England. Saltbox houses were oriented to the south with a two-story elevation fenestrated by windows. The north roof sloped down to one story and had few windows. This took the chill out of cold New England winters when the sun was out. In California, homes built in the Spanish colonial style had elongated east-west axes so that they faced the sun. Roof overhangs and decks the length of the house on the second floor shaded the south-facing glass in the summer.

As these examples show, until fairly recently we humans have been pretty good at living within our climatic means. But by the

(continued on p. 69)

Spanish colonial homes in California were built on a long east-west axis facing the sun, with roof overhangs and second-floor balconies to shade the south glass in summer.

THE CLASSIC SALTBOX

The two-story side of the house faces south for solar heat in winter, while the low north side has few to no windows and the sloping roof diverts cold north winds. The rear roof line is often extended to cover cord wood for winter heating. Shutters are operable to reduce heat loss in winter and solar overheating in summer on east and west windows.

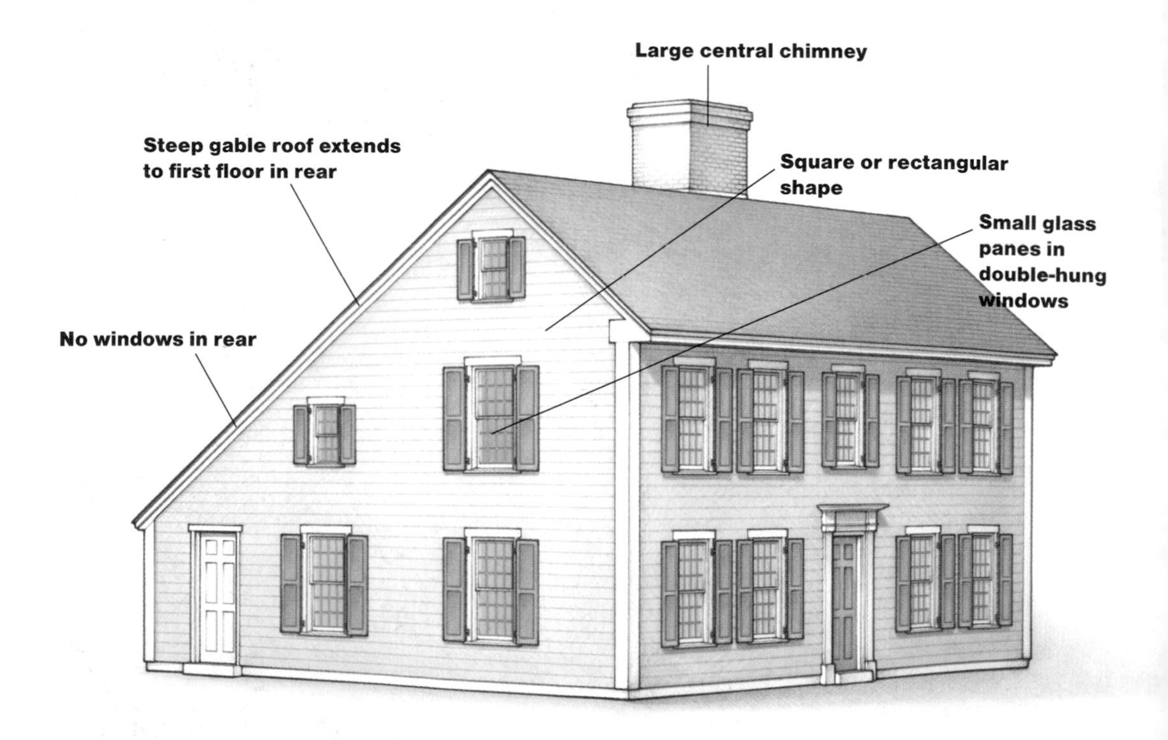

THE FATHER OF AMERICAN PASSIVE SOLAR

It's not widely known that the father of our country was an early advocate of passive solar design. George Washington built Mount Vernon in Virginia for his family after the Revolution. As a surveyor, he was very familiar with sun angles and the path of the sun over the course of the day, summer and winter. He built an attached south-facing greenhouse adjacent to the house. The floor of the greenhouse had cavities or ducts that were covered with stone. As the stone heated up, it heated the ducts and air was circulated through them and into the house.

There's little design variation in production housing across the United States. Insulation is installed to local code, and the house could be oriented with any side facing south.

(continued from p. 66)

middle of the 20th century, with the invention and commercialization of air-conditioning, we started to lose direct contact with variations in climate and weather. After thousands of years of planning shelter around the cycles of the sun, we stopped worrying about where the sun would be located in the summer and winter sky. What did it matter? Powerful heating and cooling systems took over, allowing builders to use essentially the same house designs in Anchorage, Alaska, as they did in Pensacola, Florida.

When energy was cheap, it was easy to rely on mechanical systems that overcame the conditions of nature. Now, as we are becoming painfully aware, volatile energy costs are making this approach obsolete. We need to look backward to see how people have traditionally lived as well as anticipate the environmental conditions that our buildings will have to endure in the future.

Passive Solar in the 1970s

It was another energy crisis that brought attention to passive solar design in the postwar era. When the oil embargo struck an unwary America in 1973–1974 and gas lines stretched around the block, American architects and engineers turned to the sun as one way to reduce our dependency on imported oil. In particular, Dr. Doug Balcomb, a researcher at Los Alamos National Laboratory, started exploring how to harness the sunlight that was pouring through south-facing windows in buildings in New Mexico and use it to reduce heating costs. His home, built in 1976, became almost iconic in launching a national movement to bring passive solar design to market in the late 1970s and early 1980s.

The U.S. Department of Energy launched a national passive solar program

Doug Balcomb's house was one of the first to incorporate passive solar design in the United States in the 1970s.

PASSIVE SOLAR AND ZERO ENERGY HOMES

The marriage of passive solar design with zero energy homes makes cost-effective heating and cooling possible. Solar is simultaneously the first and the last thought in a zero energy home. Orientation comes first, enabling the building to take the fullest advantage of the sun. The last thought is where and how much solar hot water and photovoltaics (PVs) to put on the building to make up the last of the loads that can't be provided by the design and operation of the home. The less solar equipment required, the more cost-effective the construction costs of the home.

in 1977. Passive Houses were built in every climatic region of the country, and advances were made every year in optimizing design. Major manufacturers of building products as well as trade associations in the building industry formed the Passive Solar Industries Council to bring passive solar to the mainstream, marking the beginning of the green building movement in the United States. But all the progress made toward a more sustainable future came grinding to a rapid halt in the mid-1980s, as political winds shifted and vast amounts of oil were discovered in Alaska and the North Sea in Europe. Oil prices plummeted, and so did interest in solar energy. The momentum was never lost in Europe, however, and the Passive House movement in Germany was born in 1996 (see chapter 1).

Passive House principles have taken root more strongly in Europe, where energy prices are considerably higher than they are in the United States.

The Design Process

Making the most of the site's solar potential is step one in the design process, coming even before any aesthetic sketches are developed. Orienting a building toward the sun, specifying the size and type of windows, and balancing light and thermal mass are all elements of passive solar design. Unlike solar hot-water systems or photovoltaic panels, which are active or mechanically controlled solar components, passive solar design relies on nothing more than good planning and design to help make the building more energy self-sufficient.

GETTING TO ZERO

Passive solar design is much easier to apply in new buildings than in existing homes. Virtually every zero energy house in colder climates has taken full advantage of passive solar heating potential; up to 60 percent of the heating load can be met through passive solar design.

Get to know the site

Begin the design process by spending time on the site. Feel the way the environment works. Every place has its own microclimate. As much as climate data will help in the macro design, the relationship to the southern sky, the available breezes, the existing vegetation, and the side of a slope the house is to be built on all create a zone that will affect the performance of a passive solar home. Use basic geometry to determine the angle of the winter sun. Is it blocked by existing trees? Do the winds wrap around a hill and hit the house from a different angle than the prevailing direction? Mature

LOCATION AND CLIMATIC CONDITIONS

Whether the house is new construction or a retrofit, local weather patterns are a basic planning tool for a zero energy home. Data should include precipitation records, average temperatures, high and low readings, and the minimum temperature by month and preferably by day over the last several years. Local airports often keep detailed weather records, and a local college or university may be able to provide projections of weather patterns and the potential for regional climate change.

In some locations, it may be apparent that both average and high temperatures have changed over the last 5 years. For example, 9 of the 10 hottest years on record have been in the last decade. Whatever the case, likely minimum and maximum temperatures are essential in sizing heating and cooling equipment, determining how much insulation is appropriate, and making a variety of other decisions about the structure.

A wind rose, which shows the direction and strength of wind at the site, can be used to design natural ventilation into the house. It is surprising that most houses are designed without regard to cross-ventilation. For many designers, windows are primarily architectural features rather than functional elements that have the potential to reduce energy use and increase comfort. As we'll see in chapter 3, deciding whether to install a wind turbine on the site to generate electricity is impossible without accurate wind records.

Rainfall records are important in two respects: first, when designing a water catchment system, and second, when determining which plants, vegetables, and fruits are appropriate for the site. Ideally, outdoor plants should be native to the area; they'll need a minimum of care and very little additional water. Moreover, backyards can be converted into gardens that provide food.

trees in the immediate vicinity can reduce the actual temperature around the house by 15°F in summer. These are not elements that will be analyzed by computer software, but they affect the quality of life in a house built on that spot.

David Barrett from Boulder, Colorado, is an experienced architect who has been designing passive solar homes for over 30 years. He camps on the site before designing every project, an experience that not only guides the house's passive solar features but also makes the architecture more harmonious with its environment.

David Barrett's work is considered organic or natural, driven by local environmental forms and microclimatic conditions.

Work with computer models

Airtight superinsulated homes perform differently than conventional homes. Many of the design teams featured in this book ran repeated computer simulations to optimize all the variables before the design was developed. For example, Jim Logan who has designed dozens of zero energy homes in Colorado, ran 15 different approaches to the Next West house, including cost–benefit analyses, to make sure he got the final design right (see "The Next West House" on p. 146).

As a specific example, the heat generated by the sun can have a rapid and significant impact on comfort, so sizing the glass is critical for both summer and winter performance. Similarly, choosing glass that functions well in all seasons, particularly east and west glass in summer, is very important. Because the energy flows are subtle, unwanted heat in hot weather can make low-energy cooling difficult. Tuning windows so that the glass has different heat-gain characteristics depending on the orientation makes the job much easier. Low solar heat gain windows for the east- and west-facing sides can reduce the heat passing through the window by as much as 80 percent. (Typical low-e glass reduces heat gain by 60 percent or more.) On the south side, it is just the opposite: High solar heat gain windows are available that allow 70 percent or more of the heat to pass through the glass, which makes a difference in climates with cloudy winters.

Choosing Glazing

Until recently, very few window manufacturers allowed you to tune your windows with respect to orientation. There were only double-glazed windows and low-e windows, with a choice of size and type. Today, the window industry is changing rapidly. Not only are customers able to specify multiple low-e coatings but multiple glazings as well. This new generation of windows, known as super-windows, has been made in Canada and Europe for years, but they haven't penetrated very far into the U.S. market.

Super-windows typically have more than two glazing layers. Windows from Canadian companies like Thermotech use fiberglass frames that are filled with closed-cell urethane foam for high R-value. The glazing is three layers of glass with low-e coating on most interior surfaces. Serious Materials, a new U.S. company, has begun producing very energy efficient windows with similar foam-filled fiberglass frames. The major difference is the glazing. Instead of three panes of glass, their windows have one or two layers of high-transparency plastic film suspended between the inner and outer panes of glass. Each of these films is also coated with low-e, and the glass is filled with inert gasses, such as argon and krypton. Serious Materials has been able to achieve

ANATOMY OF A SUPER-WINDOW

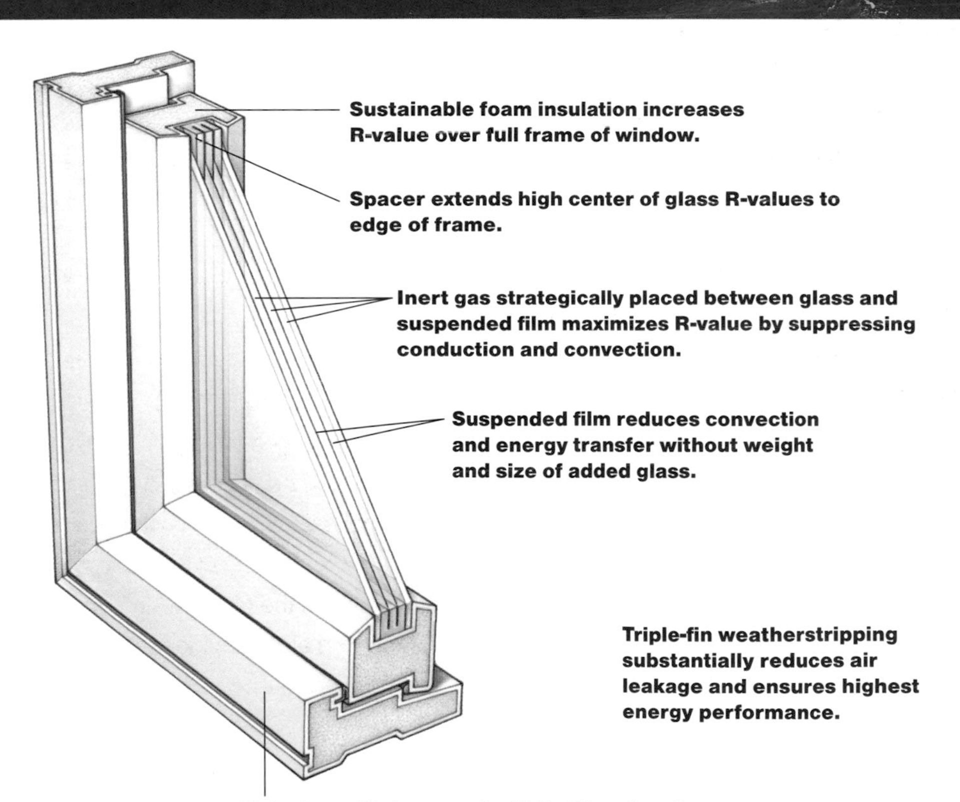

Adapted from www.seriouswindows.com

impressive R-values of 7 to 11 (U-value 0.14 to 0.09) in operable windows, which opens up a whole new world of glazing. With such low U-values and the ability to adjust other aspects of the glazing like solar heat gain, architects are able to use more glass with less of an energy penalty.

Open floor plans make it easier to distribute passive solar heat (aided by ceiling fans and mechanical ventilation) and also allow daylight to illuminate the entire space.

Thermal Mass

Thermal mass is any dense material that absorbs heat during the day and releases it at night after the sun has set. The tighter the house, the less south glass will be needed to provide comfort. Thermal mass allows for more glass and a greater passive solar contribution to winter heat and a higher performance home overall.

Mass serves a number of purposes, including absorbing light from windows and converting it to heat. Think of a warm rock on a sunny summer day. It serves as a flywheel, absorbing heat all day long and releasing it after the sun goes down. It can also absorb cool night air and keep the house temperature down in summer to reduce the need for air-conditioning through night ventilation. The ideal thickness for thermal mass in direct sunlight is 4 in., or the typical depth of a slab floor. Mass thicker than that doesn't really add much additional thermal storage. Any mass helps, but strategic placement is important. Mass that is illuminated by the sun is four times more effective than mass that absorbs reflected light only. All mass serves the purpose of stabilizing the interior environment as it warms to the average indoor temperature of the house. If the house gets too warm, the mass will absorb the excess heat and keep the median temperature lower. If the weather is cloudy for several days, backup heat of some kind is necessary.

Thermal mass can be incorporated into the design in many ways. Flooring is a great place for mass because it is the largest area of illumination. It can be as thin as ceramic tile in a thick-set base or as thick as a concrete slab.

Saltillo tile on a concrete slab is a high-performing thermal mass, holding heat and slowly releasing it when the temperature drops. The lower the windows, the more of the floor is illuminated.

The caveat is that the slab should be as close to the sunlight as possible for optimum storage. Covering the slab with carpet and padding reduces the interaction of the mass with sunlight and the house air, both during the cycle of absorbing heat and in releasing it back to the room.

THERMAL MASS HEAT STORAGE CAPACITY OF COMMON BUILDING MATERIALS

Water is the best thermal mass in terms of its thermal storage capacity. It is often used as the base reference for comparing other materials. Only the first 4 in. of thermal mass is effective in storing and releasing heat, and greater thickness yields little additional benefit. If the thermal mass is less than 4 in., divide the listed chart figure by the thickness of whatever product you are using. For example, 2 in. of builders brick would yield 3.25 Btu/°F per square foot. Hardwood flooring is included merely for comparison purposes.

Material	Btu/°F per Square Foot (4-in. depth)
Water (reference)	20.8
Brick pavers	9.0
Poured concrete	8.8
Flagstone	8.5
Concrete masonry units	8.4
Builders brick	6.5
Adobe	5.5
Hardwood	1.7

In addition to flooring, ornamental masonry or stone elements can also be used to collect heat. If a fireplace or masonry chimney is already in the plans, place it so that it can help do double duty. Mass can be placed perpendicular to the south wall running east-west so that it receives sunlight from both the southeast and the southwest. This way, it gets charged from both sides. Other approaches include using the mass for architectural features like fountains or planters.

Perpendicular mass walls have the advantage of being charged by both southeast and southwest light in the afternoon. They also provide a greater surface area to radiate heat back into the house.

Thermal mass and radiant heat

Thermal mass is often combined with in-floor radiant heat in a Gyp-Crete® (lightweight concrete) floor. Another option is to use a surface-mounted radiant heat product like Warmboard®. This is a high-density oriented strand board (OSB) that has been routed out to accept the cross-linked polyethylene (PEX) tubing used to distribute the hot water. In some installations, the entire surface is covered with aluminum to provide a greater radiant surface area. The water itself plays a significant role in storing solar heat.

Warmboard can be applied over an existing floor or serve as a subfloor. The aluminum provides even distribution of heat.

Sizing and simulation are critical if thermal mass and radiant heat are combined. If the heat source is sized inappropriately, it can overheat the house. A conventional tank water heater or an instantaneous water heater can be used for backup heat. The thermodynamics become tricky as the mass is being heated by two sources. When the mass is warmed from the in-floor heat, it can hold less heat from the sun. Heat transfer works best when the difference in temperature is greater between the solar heat and the temperature of the mass. The ideal is for the mass to cool down at night in preparation for the next morning's sunshine. A night setback thermostat that can be adjusted will allow the mass to release the available heat. It will take some trial and error for a homeowner to get the timing right. The more mass there is, the slower the process between absorption and release of heat in the mass. A surface application like Warmboard is fairly responsive, so the lag time is not very great.

Solar Shading

Passive solar works only when south-facing windows are shaded during the summer. Solar shading has been accomplished in various ways over the years. Pioneering design from both coasts used overhangs to protect the building. As previously mentioned, Spanish

Prairie-style houses as developed by Frank Lloyd Wright have wide overhangs around the entire house. The overhang on the southern side shades the glass in summer, while the overhangs on the other orientations help protect the siding from water intrusion.

colonial architecture in California used architectural overhangs to keep sun out of buildings in the hot days of summer, while in New England settlers built overhangs to protect cordwood for heating in winter. Frank Lloyd Wright's Prairie School of architecture employed wide overhangs to protect the entire building from sun and rain. The overhangs help keep water out of the walls at the top, and shading the envelope in summer helps reduce cooling loads.

Other ways to incorporate shading—awnings, louvers, photovoltaic installations, and myriad other innovative solutions—have been designed by architects from around the world. Solar shading is one of the easiest and cheapest ways to reduce cooling loads in both residential and commercial buildings. Some of the louvered devices allow for adjustment by season and local weather conditions.

Principles of Passive Solar Design

Passive solar seems simple enough conceptually. It is the balance of all the components that makes it work year round. All too often in the early 1980s, designers optimized the heating component of south-facing glass only to cause overheating and discomfort at other times of the year. Thermal mass makes a major contribution to the comfort cycle of day into night if it

THE ANGLE OF THE SUN

The sun is low in the winter sky, reaching its lowest point above the horizon on December 21. It reaches its highest point on June 21. To take full advantage of the sun, situate the house to let in the sun in winter and keep it out in summer (with appropriate shading for south-facing glass).

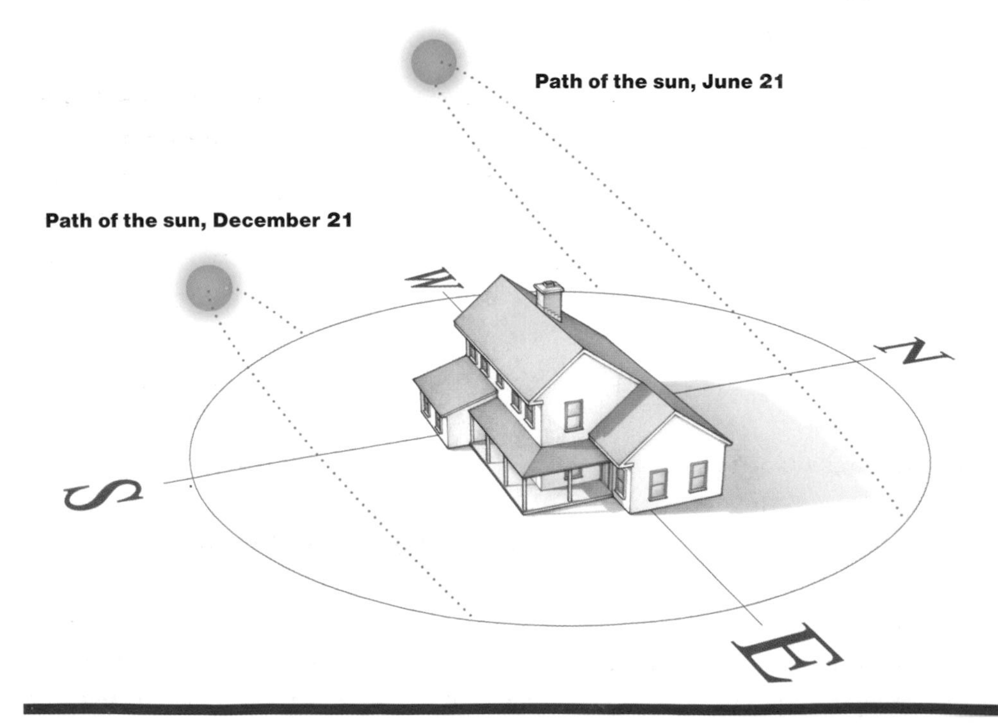

Adjustable louvers work well in milder climates where spring and fall can be either warm or cold, depending on the year. They give the homeowner the opportunity to adjust the amount of sunlight that enters the house.

is sized correctly. If it is undersized, the house temperatures can swing wildly during the day without much positive effect after the sun goes down. It is the balance of glass to mass and appropriate shading with natural ventilation strategies that makes the house work as a system. As with all aspects of a zero energy home, the components need to be integrated so the whole system works as designed.

The following list of principles should help guide the passive solar design process.

1 → Orient the house within 30 degrees of due south

Due south is the equivalent of 100 percent potential solar heat gain through windows. Rotating the house to within 30 degrees of due south still provides about 90 percent of the potential solar gain and allows latitude for adjusting the house to lot limitations. Much further than 30 degrees starts to make architectural shading difficult and can lead to overheating.

2 → Use design software to optimize passive solar heating

Several software tools are available to help optimize passive solar designs. Foremost among those is Energy-10®, developed by Dr. Doug Balcomb at the National Renewable Energy Laboratory. In its latest edition, it also assists in sizing solar hot-water systems and photovoltaics. (For more on energy modeling software, see Appendix 1 on p. 242.)

3 → Super-insulate the house

Super-insulation and air sealing (see chapter 1) are the perfect context for passive solar heating. The lower the heating and cooling loads of the house, the less solar heat is necessary to provide comfort through much of the year.

4 → Design the size of south glazing to meet the heat loss of the house

The size of south-facing glass is determined by location, the potential amount of sunshine, degree days of the local climate, the heat loss of the house, the internal architecture of the building, and intended thermal mass. In an ideal design, south windows would be distributed across the south wall to provide heat gain to as many areas of the house as possible.

5 → Configure thermal mass to absorb sunlight

Concrete, concrete masonry units (CMUs), brick, stone, and tile are typical materials used as thermal mass inside the building. The amount of glazing and insulation helps determine the amount of mass required to keep the house from overheating during the day and reduces the backup heat required during the night. Thermal mass that is directly irradiated is much more effective than thermal mass that receives only reflected light. Thermal mass requires that solar geometry be used as a design criterion for the placement of the mass.

6 → Determine appropriate overhangs for all south glass

All south-facing glass needs to be shaded in the summer to prevent overheating. To determine the appropriate size of the overhang, see "Calculating the Depth of the Overhang" below.

7 → Limit east, west, and north glass while providing for cross-ventilation

East and west windows lose as much heat as they gain. They are particularly problematic in regions with hot summers because they face the low solar angles in mornings and afternoons. That said, they provide an important function in creating cross-ventilation in rooms or for the entire house. Glazing becomes an important architectural feature, so the type of

Fixed louvers are integrated into the design of the building, allowing the perfect balance of sun into the building. As the sun gets higher in the sky, the shading extends farther down the wall so less light is let in.

CALCULATING THE DEPTH OF THE OVERHANG

Fixed architectural overhangs are designed based on the solar geometry for your latitude. The width of the overhang is determined by the solar angle of the sun on June 21 and the height of the window relative to the width of the soffit. It should totally shade the window in June and allow full solar penetration through the window on December 21.

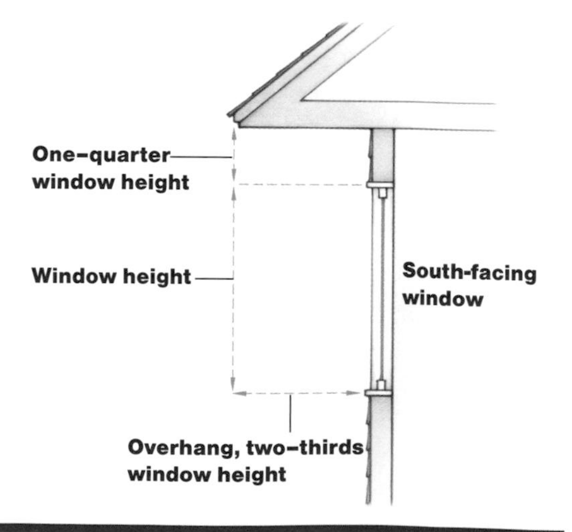

Trees planted on the northwest side of this house block hot afternoon heat in summer and protect the house from the prevailing winds in winter.

GETTING TO ZERO

Designing a house to respond to its site and climate is the first prerequisite of zero energy building. Finding ways to heat, cool, and light the space with the barest amounts of energy is the next. Passive solar is the biggest bang for your buck in providing comfort. Natural daylight also reduces electrical loads during the day. The themes introduced in this chapter will crop up repeatedly, in slightly different forms, in the houses that are explored throughout the book.

glass is important when placing east-west windows. Low solar heat gain glass can reduce the potential for overheating.

8 → Design appropriate shading strategies for east and west glass

The easiest way to limit summer heat gain through east and west windows is through strategic planting outside. Trees and bushes that leaf out in summer and drop their leaves in fall provide for light and ventilation while shading the window when it is hottest outside. Be careful that the species selected doesn't grow too tall and lose its lower branches, thereby reducing its shading potential.

9 → Calculate the backup heating and cooling required

In a house insulated to current codes, thoughtful passive solar design can reduce the cost of heating and cooling by as much as 50 percent. In a superinsulated house, passive solar can replace even more fossil fuel than that. Good design is the key. Overheating is one of the main reasons that builders are skeptical of passive solar. Design software has to be adjusted to take into account the passive solar contribution before sizing HVAC equipment. Typically, the heating load is so low that simple heat sources, like a conventional water heater for radiant heating, are sufficient. Split-system heat pumps can work for many parts of the country.

Case Study

PassivHaus Retrofit

Creating a near zero home from a 1904 Craftsman bungalow

It's one thing to build a new zero energy house, but quite another to retrofit an older house to meet contemporary demands for energy efficiency and livability. Much of the housing stock in Berkeley, California, is old classic bungalows and Craftsman designs that were built long before energy was an issue. In recent years, Berkeley has put itself on the map not only for its unique character but also as a leader in environmental policies.

SPECIFICATIONS

House size: 1,515 sq. ft.

Average heating degree days: Approx. 2,600

Wall and roof construction: First floor (new), 2×6 at 24 in. o.c.; second floor (existing), 2×4 at 16 in. o.c.

Insulation type: Cellulose throughout, with additional 2-in. rigid insulation on exterior of second-floor walls

R-values: Foundation, R-19 above slab on grade (rigid $1^1/_2$-in. polyisocyanurate); walls, first floor R-20, second floor, R-26; attic floor, R-38 (blown cellulose)

Windows: Sierra Pacific, casement, wood with aluminum cladding

Photovoltaic capacity: Not available

Heating source: Electric baseboard heaters; energy recovery ventilator (UltimateAir®)

Air-conditioning: None

Cost per square foot: Not available

Performance data: Designed to Passive House standards, which has a maximum heating or cooling energy of 1.4kWh/sq. ft./year

Architect: Nabih Tahan, AIA, MRIAI

General contractor: Christopher Polk

Home performance contractor: George Nesbitt

This simple yet beautiful 1904 home uses considerably less energy than any of its neighbors (just 1.4kWh per square foot per year).

In 1976, Architect Nabih Tahan bought a 1904 Craftsman bungalow in a great neighborhood, close to schools and public transportation, but the house was cold and drafty and needed a lot of work. He spent two years remodeling the home, and became the first person in the United States to retrofit a home with the goal of meeting the PassivHaus standards.

APPLYING PASSIVHAUS PRINCIPLES TO A ZERO ENERGY HOME

Tahan was very familiar with PassivHaus principles as he had worked for years in Europe designing low-energy homes. This method of building stresses the importance of a super-tight, well-insulated thermal envelope and uses very low amounts of fossil energy to heat the home. In fact, Passive Houses rely mostly on the free heat given off by the occupants, appliances, and equipment to heat the home.

The Passive House standards state that the heating requirement cannot exceed 1.4kWh per square foot per year. This is an ambitious goal, but by investing in the home's envelope and paying extra attention to airtightness, Tahan is confident that his home will meet the goal. Typical Passive Houses don't have a big heating system and hence eliminate the usual costs of installation, upkeep, and operational costs of conventional mechanical systems.

Raising the House

For Tahan, the road to zero energy began in the basement. The original first floor had only 7 ft. headroom, and the existing brick foundation was damaged and in need of replacement. Tahan had to raise the house up 3 ft. to achieve legal floor-to-ceiling height. He also raised the slab about 1 ft. above grade.

Insulation is key to the PassivHaus standards. Although Berkeley has a mild climate, it was still important to insulate the foundation to prevent heat loss. A new slab was poured with 1-in. extruded rigid polystyrene insulation (R-5 per inch) below the slab and on the exterior of the stem wall. Two layers of 1 1/2-in. polyisocyanurate rigid foam (R-9.8 per layer) were installed on top of the slab between two layers of 2×2 sleepers to totally isolate it from the ground. The subfloor was attached to the 2×2s.

Insulating to PassivHaus standards

Tahan used calculations from the Passive House Planning Package (PHPP) to determine R-values for each floor's framing. The walls of the new first floor are built with advanced framing techniques using Forest Stewardship Council–certified 2×6s at 24 in. o.c. with blown-in cellulose insulation (R-20).

Buying products locally cuts back on energy spent in transportation. These Sierra Pacific double-pane casement wood windows with aluminum cladding are manufactured in California. The existing redwood siding on the house was removed, refinished, and reinstalled.

Two layers of 1 1/2-in. polyisocyanurate rigid foam on top of the new slab between 2×2 sleepers totally isolate the slab from the ground.

Air sealing is critical to prevent air loss and moisture movement through the walls. The sheathing is sealed with caulk to provide the air barrier. A blower-door test will show if any air leakage has been missed.

As in other PassivHaus buildings, airtightness was critically important, and no penetrations were left unsealed. The plywood sheathing was carefully caulked at the joint between sheets to form the air barrier, as Tahan didn't feel that the standard building wrap was sufficiently airtight. The second floor used the existing 2×4 construction with new blown-in cellulose insulation. The plywood sheathing was then wrapped with 2 in. of polyisocynurate rigid foam for an R-26 wall assembly. The windows are double-pane low-e (rather than triple-pane, given the Berkeley climate). All the interior walls were also blown with cellulose to reduce sound transfer from room to room.

Advanced air-sealing techniques were used at all intersections of the building planes: where walls come together and at floor-to-wall and wall-to-ceiling intersections. Expandable foam was used for all penetrations such as hose bibs and the electrical service entrance. Foam was also used in interior walls where pipes and wires penetrate framing plates from one floor to another. In the ceiling, Tahan blew 10 in. of dry cellulose between the new 2×10 joists. The attic access is an insulated hatch made by Rainbow Attic Stair, with 2 in. of Styrofoam®, which has an R-value of 10.

Expandable foam was used to seal up gaps at the ceiling-to-roof intersection in the attic.

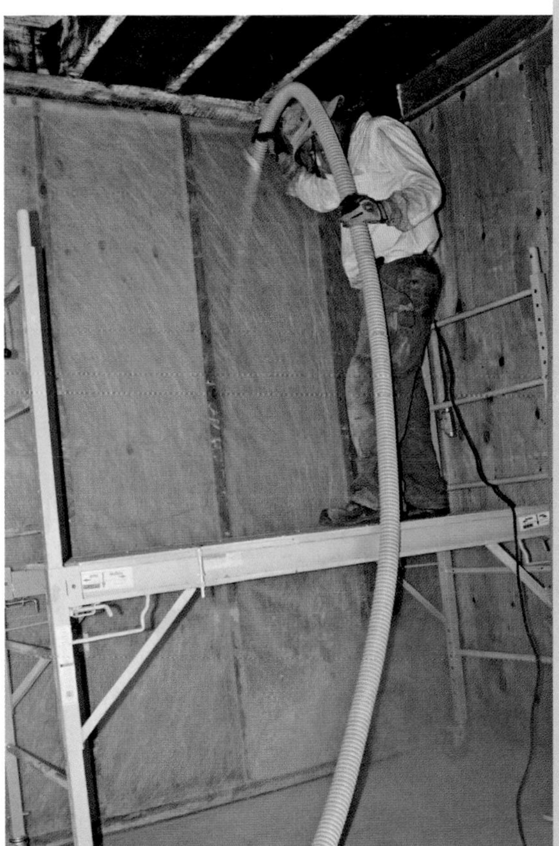

Dry cellulose is an excellent insulator but not a good air barrier. Less expensive than urethane foam, it is made of 80 percent post-consumer recycled newsprint.

GETTING TO ZERO

Architect Nabih Tahan worked to optimize each part of the envelope of his house. "Measuring energy is very important," says Tahan. "If we want to lose weight, we count calories; if we want to measure a house's performance, we count kilowatts. In Europe, hundreds of houses are monitored for energy performance to better understand exactly how more energy can be saved and what building methods work best."

Expandable foam was also used to seal up gaps around pipes.

A Rain Screen Allows Moisture to Escape

In the Bay Area, rain from the Pacific in winter can come down in horizontal sheets that drive moisture into and behind conventional exterior wall finishes. To protect the carefully designed wall structure, a rain screen was installed under the final stucco finish, creating an air layer between the house wrap and the finish. The rain screen is formed from 1×2 battens nailed over the housewrap and directly into the studs. Cement board was installed over the battens as a base for the stucco. This assembly allows any moisture that gets behind the cement board to flow down the wall and out through weep holes at the bottom. On the second floor, the original redwood siding was carefully removed, flipped over, refinished, and stained to maintain the patina of the Craftsman style. The boards were also attached to battens over the housewrap.

Energy-Recovery Ventilation Helps Keep the Air Clean

By preconditioning the air entering the house, removing harmful pollutants and moisture, filtering out pollens and molds, and providing fresh filtered air, energy-recovery ventilation (ERV) technology creates a healthier indoor environment year round. An equally compelling reason for using ERV is the fact that it conserves energy and reduces peak demand on cooling systems.

Tahan installed an UltimateAir ERV to supply fresh air in the sealed envelope and to maximize heat retention. The heat that is generated by people, appliances, lighting, and equipment is captured by the ERV and recirculated back into the house through a heat

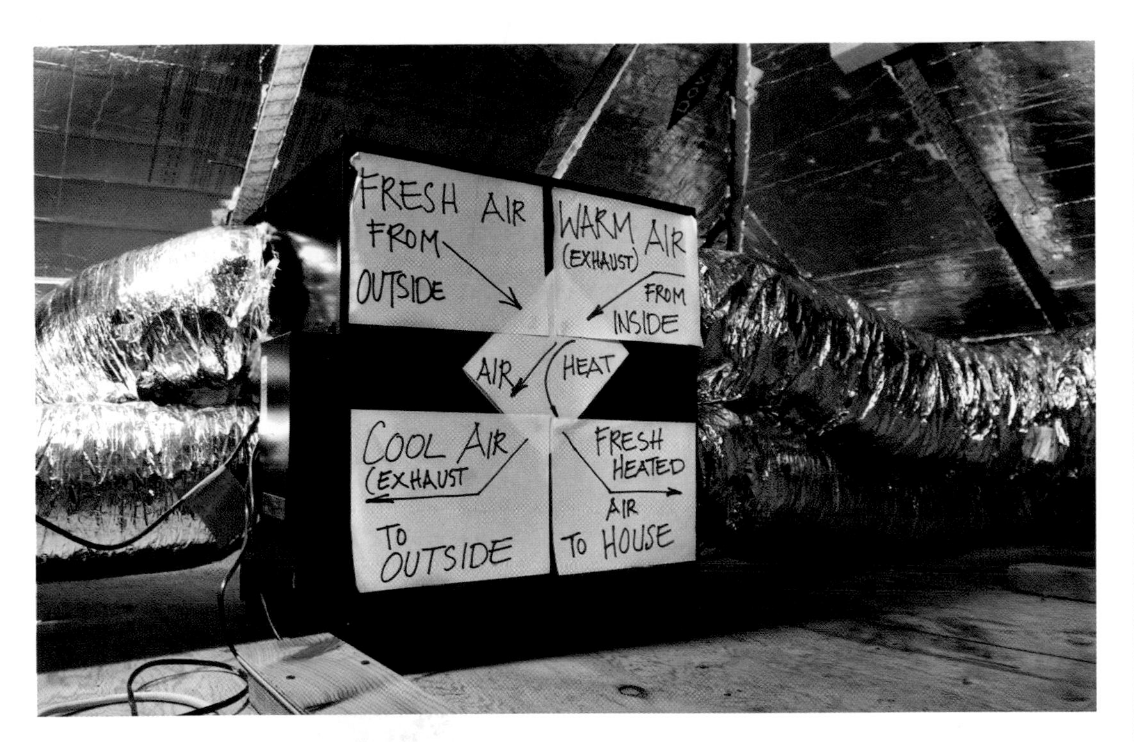

An ERV unit saves energy by exhausting stale air while simultaneously introducing fresh air into the home.

exchanger. A dehumidistat, which measures the moisture content, was installed in the bathrooms and kitchen. Once the humidity reaches 60 percent, the ERV is kicked up to high speed to exhaust the air faster from these rooms. The heat in the exhaust air is recycled by transferring it to the fresh incoming air.

During periodic maintenance and cleaning of the filters for the ERV, Tahan realized that the wheel to the ERV was unfortunately stuck in place. The wheel, when spinning, transfers the heat from the outgoing air to the incoming air. This has skewed the energy performance data for the first year, which should be more accurate for the second and subsequent years. Even with this unfortunate mishap, Tahan believes that the energy consumed for heating is in line with Passive House requirements.

Being a pioneer is never easy. The PassivHaus concept is still very new in the United States, particularly for a remodel. (It is always easier to start from scratch when building a zero energy home.) Remodeling has its complications, which made this project a great learning exercise. Much work has yet to be done coordinating with building departments across the country. The International Building Code has made improvements in how it covers green building, but very energy efficient projects are still beyond its scope. The proof is in the final product, and this home is being used as a demonstration building for architects and builders across the Bay Area, showing what can be done to convert an existing house to zero energy.

Case Study

Bringing 1887 into the 21st Century

A near zero infill in downtown Santa Barbara

Property in downtown Santa Barbara, California, can be pricey, and houses in its historic district must fit in with prevailing architectural patterns. Two earlier development plans for the lot that long-time contractor Dennis Allen bought here had been rejected by the city. But Allen's ideas and his standing in the community finally helped him win approval for a near zero energy project.

SPECIFICATIONS

FRONT HOUSE

House size: 2,500 sq. ft.

Average heating degree days at site: 2,283

Wall and roof construction: 2×6 wood framing, partial truss roof

Insulation type: Open-cell spray-in urethane

R-values: Foundation, R-30; walls, R-21; roof, R-30

Windows: Double-glazed, low-e

Wind/photovoltaic capacity: 4kW

Heating source: Modulating gas, forced air

Air-conditioning: None

Cost per square foot: $400

General contractor: Allen Associates

BACK HOUSE

House size: 3 units totaling 4,860 sq. ft.

Average heating degree days at site: 2,283

Wall and roof construction: 2×6 wood framing, structural insulated panels on roof

Insulation type: Open-cell urethane foam, rigid foam

R-values: Foundation, R-30; walls, R-24.5; roof, R-30

Windows: Double-glazed, low-e

Wind/photovoltaic capacity: 6kW

Heating source: Gas-fired tankless water heater with forced-air distribution; water preheated by solar panels

Air-conditioning: None

Cost per square foot: Unknown, project ongoing

General contractor: Allen Associates

This CAD drawing shows what the reconstructed Victorian façade in the front and the new condominium behind it will look like when completed in late 2010. The building retains the character of the neighborhood but reduces energy consumption drastically.

Originally, Allen had hoped to renovate the 1887 house on the property and transform it into a zero energy project. But as he got into the bones of the house, it became clear that renovation would not be possible. Framing throughout the house had rotted, and the floor sagged 7 in. from one side of the building to the other. In addition, walls framed with 2×3s left so little room for insulation that conversion to a zero energy home was virtually impossible.

A Modern Building in History's Disguise

As a compromise, the city's Historic Landmarks Commission agreed to a reconstruction of the historic façade with a virtually new building behind it. In the back, Allen planned to build a new, three-unit condominium building.

What made the project possible (and pleased the city) was the use of auto lifts to double the number of cars that could be parked on site. A parking elevator allows one

The façade of the existing Victorian house was saved, but much of the framing had rotted away and become unsafe, requiring an entirely new building that still matched the existing detailing.

car to enter the garage and be elevated so that a second car can park underneath it. Instead of dedicating space on the project for eight parking spaces, there are only four, thus making more room for a community garden and improved solar access to the rear building.

Both buildings are designed to be near zero energy. The existing home was taken apart, with materials still fit for service put aside. Siding, for example, was remilled and used on the front façade of the reconstructed home.

The existing siding was remilled and used again on the front façade. Reusing building materials cuts down on landfill waste and helps preserve the historic charm of a building.

Allen worked with a mechanical engineer who had helped develop energy modeling software at MIT along with students from the University of California at Santa Barbara. The team ran 15 computer simulations with various energy scenarios before selecting basic systems for both buildings, focusing efforts on the back building, for which they had more design flexibility. Computer simulations analyzed both energy performance and the cost and benefits of each option. (For more on energy modeling software, see Appendix 1 on p. 242.)

Framing to Meet Seismic Codes

The original design was to use structural insulated panels (SIPs) on the walls of both buildings, but all SIP construction in Santa Barbara had been put on hold by the state because of seismic considerations. (SIPs were, however, used for the roof.) In the end, what worked best was 2×6 wall framing insulated with Icynene®, an open-cell spray urethane insulation. Framing materials were certified by the Forest Stewardship Council. Sheathing was wrapped with DuPont® Tyvek® ThermaWrap™, which added R-2 to the walls and provided a thermal break.

DuPont Tyvek ThermaWrap has high vapor permeability and reduces radiant heat flow through the wall system.

Icynene open-cell spray foam is an excellent air barrier and insulator. It does not shrink, sag, or settle and minimizes air leakage for increased energy efficiency. Spraying into the sloped roof provides conditioned space in the attic for ductwork.

On the rear building, 1/2-in. rigid foam insulation covers the sheathing. A drainage plane was created over the insulation with primed and sealed 3/8-in. CDX plywood strips nailed vertically to the studs before the siding was applied. This provides a continuous air space from the top of the wall to the bottom to allow any moisture that gets past the siding to drain down and away from the building. Weep holes at the bottom of the siding sealed with an insect screen protect the structure from wood-boring insects.

Loewen low-e squared windows were installed in the front house to meet Historic Landmarks Commission requirements. Therma-Proof windows, manufactured by Serious-Windows™, were used on the rear building and are rated at U-0.15 (R-7).

Battens were used to form a drainage plane behind the siding. By allowing moisture to escape, mold and decay are much less likely to be a problem.

Let the Sun Shine In

Dennis Allen has been building energy-efficient, passive solar buildings with very low heating loads for more than 30 years (in Santa Barbara there are no cooling loads to speak of). The top two floors of the rear building employ a passive solar strategy, using direct-gain solar heating through south-facing windows. Shades prevent overheating in the warmer seasons, and thermal mass is provided by a double layer of 5/8-in. drywall and floor tile set in a mortar bed. The passive solar design kept construction crews warm all winter without any additional heat in the buildings.

Loewen low-e windows were used on the front building to achieve a historic look. Low-e windows are coated to help keep summer sun out by reflecting more heat than conventional glass.

The combination of good passive solar design and a tight thermal envelope allowed Allen to use smaller heating units in both buildings. A York® 97 percent efficient modulating gas furnace with heat-recovery ventilation (HRV) was used in the front building, while Rinnai® tankless water heaters with fan coils were used for backup heating in the rear units. Because the buildings are so tight, the HRV was necessary for fresh air ventilation.

Invisible Solar Panels

Both buildings incorporate solar hot-water and photovoltaic panels for generating electricity. Because the Historic Landmarks Commission wouldn't pass the project if solar panels were

ThermaProof windows were installed on the rear building. These windows, made in the United States, have a much lower U-value than almost any other window on the market.

visible, they are hidden behind a parapet wall on the rooftop.

The solar hot-water system is passive. Inside the CopperHeart integral hot-water collectors are 4-in.-dia. copper pipes in which water circulates via convection and is stored in an integral tank next to the collector. There are no moving parts, so the system is virtually maintenance free. As noted, instant water heaters are used for backup.

After studying electrical loads, designers called for a 2.4kW PV system in the front building; the capacity of the building in the rear is 4.6kW. To reduce electrical loads, designers specified EnergyStar appliances, including an induction cooktop. Lighting will be primarily LED with some fluorescent fixtures.

California's Title 24 is the most stringent energy code in the country. Allen's multifamily building at the rear of the property is built to beat Title 24 requirements by 90 percent; in the front building, performance is expected to be slightly lower than that due to historic constraints. Both buildings are designed to Leadership in Energy and Environmental Design (LEED) Platinum standards.

So how much more does a zero energy building cost? Beating Title 24 by such a wide margin added only 1/2 percent to the cost of Allen's project versus the original estimate designed to meet code minimums. The buildings are scheduled for completion by late 2010.

LESSONS LEARNED

Every project holds some lessons. Here's what project manager Bryan Henson picked up on this near zero project in Santa Barbara:

- **The time and money spent modeling the building were well worth it. Multiple modeling runs not only helped us reach our passive energy goals but also saved money because we didn't overinsulate.**
- **We added a rain screen after design had been completed. The air channel made all of the architectural details and reveals a challenge. A little more planning would have sped up the process.**
- **A little trade contractor training goes a long way. Our recycling program and waste reductions strategies could not have happened without their help.**
- **Building the team between architect, contractor, owner, mechanical engineer, and key subs makes for a higher-quality, higher-performing, lower-cost building through coordination, communication, and teamwork. It's the only way to go.**
- **Precut floors, walls, and roof packages are faster, more material efficient, and entirely possible with good planning.**
- **Concrete with 50 percent fly ash isn't as scary as people might think. It's stronger and doesn't cost any more.**
- **Low-VOC products make a drastic difference. It's impressive to be in a home being painted by a team of painters and not smell a thing.**
- **Reused materials are sometimes competitive with new materials, but they have a story to go with them. We used recycled barn beams for flooring, old cedar water storage tanks for decking, and recycled beams for architectural beams. The flooring and decking cost the same as new material, but the wood is better, straighter, and reused.**
- **The passive design meant we could significantly downsize our heating system.**

Case Study

Target Zero House in Taos

A German Passive House in the Southwest incorporates airtight design

Taos, New Mexico, has long been a center for building innovation, and passive solar houses have been right at home there since the 1970s. While Taos is known for its hot summers, it also has long and cold winters with 6,800 heating degree days a year. Joaquin Karcher, an eco-designer with an architectural degree from Germany, has spent the last three years studying and developing the German Passive House methods for the cold climate of the high-desert Southwest. The result is an energy standard that allows an amazing 90 percent reduction in typical space-conditioning needs.

SPECIFICATIONS

House size: 2,200 sq. ft.

Average heating degree days: 6,800

Average cooling degree days: 222

Wall and roof construction: TJI® framing with dense-pack cellulose; viga ceiling and roof framing

Insulation type: Cellulose

R-values: Foundation, R-31; walls, R-48; roof, R-70

Windows: Sierra Pacific with Cardinal LoE^3-366®

Photovoltaic capacity: None

Heating source: Passive solar, internal gains, solar hydronic

Air-conditioning: none (HRV system, whole house ventilation system with 95 percent heat recovery)

Cost per square foot: $250

Designer: Joaquin Karcher, Zero E Design

General contractor: Owner/builder

The one-story house, built in the Southwest style with a stucco finish, blends in beautifully with its surroundings.

Karcher's 2,200-sq.-ft. house, built in 2008, is extremely airtight and has a whole house ventilation system with 95 percent heat recovery. He calls the house a *target zero home,* which is about 80 percent more efficient than a conventional home in the area. Typical net zero homes are clad with PV systems, but Karcher feels that putting a 10kW system on a roof is a ridiculous expense. "If net zero isn't affordable, it's not worth it. If it can't be done at a mass scale, then we have failed."

The Key Is to Drive down the Load

Karcher's approach is that, in design, conservation needs to come first. As he says, "The whole house has to function as a system and no part of the system can be neglected. The cheapest gallon of fuel is the one that you don't need." The biggest mistake typically made with net zero homes is to focus on the energy supply side. Instead, according to Karcher, the first step should be to drive the load down to almost nothing and only then look at the supply side. Energy modeling is critical to a good project. Rules of thumb and experience ratios are no longer sufficient at this level of performance. This project used the Passive House Planning Package (PHPP) program (see p. 18) to design the house.

The home is built on a foundation that has 9 in. of insulation under and around the concrete. The exterior walls are made from extra-thick 12-in. TJI (wood I-joists). This superinsulated envelope has an R-value of 48 and virtually no air infiltration, and advanced framing techniques helped cut down on lumber use and thermal bridging.

Karcher believes that it's relatively easy to save half, even three quarters, of the energy usage of a typical home. The difficult part is that every percentage point beyond that becomes increasingly harder to achieve. The home has a HERS rating of 28—without the use of any PV (see "What Is a HERS Rating?" on p. 100). With a 2kW to 3kW system, the house would be close to a HERS 0 rating. The super-efficient home has four 4×8 Viessmann® solar hot-water collectors tied to a radiant-floor heating system that meets nearly all of the energy demand. A propane backup (required by code) comes on only when a succession of cloudy days slows the production of solar hot water. It is expected

The superinsulated foundation is rated R-31, which is higher than the walls of most homes.

The thick walls are framed with 12-in. I-joists, which are more typically used in floors or roofs. Here, they provide ample room for insulation.

to use about $250 in propane a year at current prices.

Although this home has significantly reduced the homeowners' energy bills, that wasn't the only reason for building a near zero home. According to owner Frank Oatman, "It wasn't just energy conservation. We're both life-long conservationists. Our idea was that we thought it was morally the right thing to do for the Earth." Oatman thought it was crazy not to build a house using the sun in New Mexico, and he wanted to build a super-energy-efficient house because it was a very practical solution that would save a lot of money in the long run. And he was interested in the idea that the house could be used to encourage other people to do the same thing.

The airtightness of the home was tested using a blower door test. The house scored an excellent 1.2 air exchanges per hour at 50 pascals of pressure (ACH50).

Sunlight creates a warm and inviting feeling in this traditional, Southwest-style kitchen.

WHAT IS A HERS RATING?

A home energy rating system (HERS) rating is an analysis of a home's projected energy efficiency in comparison to a "reference home," based on the 2006 International Energy Conservation Code. A HERS rating involves both an analysis of a home's construction plans, as well as on-site inspections and testing by a certified home energy rater. Tests typically include a blower-door test (to test the leakiness of the house) and a duct test (to test the leakiness of the ducts).

The lower a home's HERS Index, the more energy efficient it is. A home built to current code should score a HERS Index of approximately 100, whereas a net zero energy home should have a HERS Index of 0. Each 1-point decrease in the HERS Index corresponds to a 1 percent reduction in energy consumption compared to the HERS reference home. Thus a home with a HERS Index of 85 is 15 percent more energy efficient than the reference home, and a home with a HERS Index of 80 is 20 percent more energy efficient.

Getting the Balance Right

Living in a net zero energy home can be more complicated than living in a regular home, according to the homeowners. The systems can be somewhat alien to someone whose previous interactions with energy delivery amounted to adjusting a thermostat on the wall, phoning the propane dealer when the tank got low, and hauling in another load of wood. "We insisted on learning all about the mechanical systems," Oatman said. "It took a while to get the hot-water system all balanced. You have to watch and make sure the pressure stays right on the system. There are several pressure gauges, and we keep our eyes on that. We pretty much understand how the system works, but it's kind of complicated."

The logs in this traditional ceiling carry 2×8 roof joists above.

While Karcher is proud of this and his other target zero home projects, he feels that the only way the idea is going to really catch on is through bold and significant code changes. If there is enough political will, there will be a market. Karcher has seen a significant shift in public opinion that is moving away from seeing green as something exotic and instead as something mainstream. Awareness about our energy use and needs is rising. Karcher thinks that EnergyStar needs to drastically raise their standards and that manufacturers will then follow. His two upcoming projects, one in Santa Fe and one in Taos, will evolve this standard and push the envelope of performance even further.

GETTING TO ZERO

New Mexico is looking to adopt the 2030 Challenge, which is an initiative that challenges the global architecture and construction community to adopt a series of greenhouse gas reduction targets for construction. Today, all current construction and remodeling projects need to reduce the building's energy use by half relative to current codes. The reduction goes up by 10 percent every 5 years, until, in theory, all buildings in the United States are carbon neutral by 2030.

Typical custom homes in Taos cost around $250 to $300 per square foot. This project came in at approximately $250 per square foot, including all the high-performance features. The cost savings were due to diligent selection of cost-effective methods and materials.

3 Renewable Energy

Zero energy houses are unique in their capacity to produce as much energy as they consume. Even the most energy-efficient house on the planet can't reach net zero without this ability to balance the energy ledger and make up for the power used for lights, heating and cooling, and appliances. And that takes a source of renewable energy.

There are two practical energy sources for most homeowners: sun and wind. For a variety of reasons, photovoltaic (PV) panels are the most common choice for producing electricity on a residential scale. The network of suppliers and installers has blossomed, and solar panels can be installed on houses where a wind turbine would be impractical if not impossible. Even so, wind turbines can make electricity less expensively than PV and under some circumstances would make a better choice.

Water runs a distant third, although it shouldn't be discounted completely. Microhydroelectric generators are a possibility for those blessed with some mechanical ingenuity and streams in their backyard, but residential-size hydro is not a realistic option for most homeowners.

Solar hot-water collectors are often part of a renewable energy package, too, although not all net zero houses include them. Several

A photovoltaic panel tilted toward the sun produces completely clean electricity and should continue to operate for 25 years or more. Continued improvements in efficiency, along with more aggressive government subsidies, are making solar electricity a more viable part of the residential energy mix.

types of collectors are available, making them suitable for just about any climate. In the right area, solar collectors have the potential to supply most of a household's hot-water needs, and they may even produce enough to help out with a hydronic heating system. Climate, geography, and household habits all play major roles in determining how much of the hot-water supply collectors can provide.

Ultimately, the choice of which form of renewable energy makes the most sense comes down to a variety of factors: electrical loads, climate, local ordinances, site conditions, subsidies from government and utilities, and, of course, cost.

Even though relatively little residential electricity or hot water is currently generated with renewable energy systems, there is enormous potential for growth and the greater energy self-sufficiency that will bring. Under conventional thinking, the challenge has been capturing solar or wind energy at a cost that's competitive with fossil fuels or electricity distributed over the grid. But because we can't look into our crystal balls to determine what natural gas prices or electricity prices will be in the future, calculating "payback" is impossible. A more realistic way to look at the value of renewable energy is to calculate the incremental cost of the system in a home loan and determine what the monthly payment will be. If the system reduces utility bills more than the carrying cost of the loan, it has an immediate positive return on investment. This return will grow only more attractive as energy costs rise.

GETTING TO ZERO

The solar energy reaching the surface of the Earth every hour is greater than the total amount of energy the world uses in a year, according to the U.S. Department of Energy. The trick is capturing it.

From Sunlight Alone, Clean Electricity

Photovoltaic cells generate electricity directly from sunlight. For nonscientists, it's a wonder how photovoltaic components work and yet somehow they do. The first truly useful photovoltaic panels were manufactured more than a half-century ago by Bell Labs, and since then the technology has seen steady progress with cells that convert sunlight to electricity ever more efficiently.

Photovoltaic panels that generate electricity directly from sunlight are typically part of a net zero strategy. The cost of PV-generated power has steadily declined, but it's still generally more expensive than utility-provided electricity.

MAKING POWER FROM THE SUN

Photovoltaic cells are solid-state devices (that is, no moving parts) that generate an electric current when struck by sunlight. When a photon of light is absorbed by material inside the cell it dislodges an electron, creating an electric current. According to the government's energy office, the photovoltaic effect was observed more than 100 years ago. But it wasn't until the 1950s that cells could produce enough electricity to be useful.

Nearly all PV cells manufactured today incorporate crystalline silicon, a light-absorbing semiconductor. Monocrystalline silicon wafers are cut from a single crystal; multicrystalline silicon is made by cutting into a cast block of silicon. Of the two, monocrystalline silicon PV cells are somewhat more efficient.

More recently, manufacturers have started to make thin-film solar cells from amorphous silicon. The material is applied in a very thin film on glass, stainless steel, or another substrate. Because thin-film cells use less silicon, they're cheaper to manufacture but also have efficiencies of about 8 percent, which is lower than the best monocystalline cells. Materials other than silicon can be used to manufacture cells as well, including cadmium telluride and copper indium diselenide.

Photovoltaic cells can be graded by their efficiency, which is the proportion of potential solar energy that is converted into electricity. The most efficient cells now commercially available are roughly 20 percent efficient.

Researchers in public and private labs all over the world are looking for ways to make more efficient and less expensive cells. One avenue of research is to look for cheaper alternatives to silicon, which is under heavy demand by the electronics industry. Another is to develop cells that shed more than one electron for each photon of light from the sun, which would increase the electrical current.

The research seems exotic to nonscientists, but lower costs and greater efficiencies hold tremendous importance for getting photovoltaic panels on top of more rooftops. If the average cost of PV electricity was $2 or $3 per installed watt instead of $8 or $9, a lot more people could afford to replace grid power completely.

Old and new: A modern crystalline silicon solar cell and, in the plastic case, the first solar cell created by Bell Labs in the 1950s.

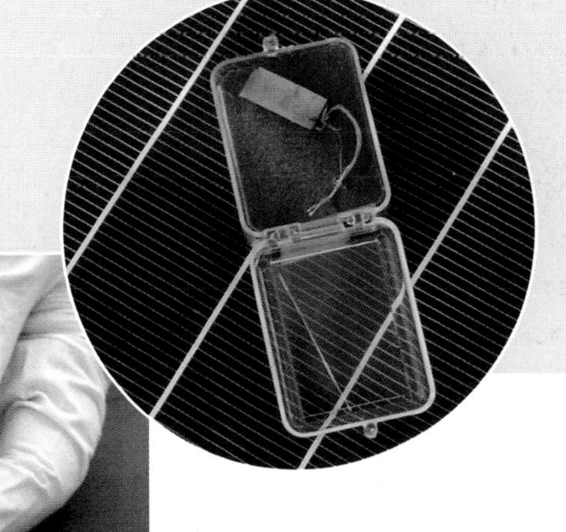

Thin-film laminates can be applied to roofing materials as a flexible membrane in building-integrated PV installations.

GETTING TO ZERO

Costs keep coming down. The cost of photovoltaic panels has declined by about 4 percent a year for the last 15 years, according to Solarbuzz™, a solar research and consulting company.

PV also is attractive from the standpoint of its energy payback, which is the time it takes for a photovoltaic system to generate an amount of energy equal to what was required to manufacture it. According to government studies, the payback period ranges from 2.7 years to as little as 1 year, depending on the type of cell. Assuming that a cell will last 30 years and will still be putting out 80 percent of its maximum rated power at the end of that period, nearly all of its productive life is on the positive side of the ledger.

Most net zero PV systems are *grid-tied,* meaning they are connected to the electric utility's grid and can import electricity as well as export it. This two-way exchange of electricity, called *net metering,* is essential because it gives the PV system a place to store an unlimited amount of excess electricity, although rates paid by utilities vary.

In contrast, off-the-grid systems with no utility connection store power in banks of deep-cycle batteries. This makes the system more expensive, less efficient, and more complicated

(continued on p. 109)

A grid-tied photovoltaic system uses a two-way meter that clocks outgoing and incoming electricity. At the end of the year, an all-electric net zero house will show a zero balance. Mixing fuels makes the calculations more complex.

BUY-BACK RATES ARE KEY FOR NET METERING

Most zero energy houses take advantage of net metering, a system in which excess power produced by a residential renewable energy system flows into the utility grid. Net metering is on the books in most states, although rules vary widely.

At certain times of the day, or in certain seasons, photovoltaic panels or wind generators may produce more electricity than the house consumes, so the meter essentially runs backward as this power is diverted into the grid. At times when the house can't provide as much power as its occupants need, the grid supplies the power.

The ebb and flow of electricity for a grid-tied house is what makes it possible to get to net zero. But there is no standardization on how utilities in net-metering states reimburse residential providers or how long credited electricity will be carried on the books.

Feed-in tariffs

When the utility is buying power at the same rate that it charges a retail customer, the cost of installing enough photovoltaic panels can still be prohibitive to many homeowners. But what if electricity generated by a house was worth two or three times as much per kilowatt hour as power generated by an electric utility? That is, every kilowatt hour produced on someone's roof and exported to the grid was worth several times as much as each kilowatt hour that homeowner had to buy. That's the idea behind feed-in tariffs, a system that has worked wonders in increasing the amount of renewable energy in Europe and Canada and is being proposed in various parts of the United States. In Germany, a feed-in tariff has helped create 20,000MW of new wind capacity, almost half of that locally owned, and 2,500MW of on-site solar capacity.

The concept is that utilities pay more (up to 10 times as much) for their electricity during peak load periods. A peak load is the time there is most demand for electricity from the utility. Imagine a July afternoon in Atlanta when air-conditioning is running at full tilt and businesses are using office lights, computers, printers, and copiers. The utility must have enough capacity to meet peak loads to avoid brownouts and blackouts. It turns out that PV produces its highest output during peak load periods for most utilities so the avoided cost for utilities is significant.

Because the cost of expanding renewable energy resources is spread over the utility's entire rate base, the argument is that the burden of feed-in tariffs on an individual rate payer isn't very much. Program specifics can be complicated, and rates may vary, depending on the source of the renewable energy to favor small producers and forestall windfall profits from big ones. For example, under a proposal in Michigan, electricity generated by photovoltaics would be worth 65 cents per kilowatt hour but 10 cents per kilowatt hour for hydro.

Indirectly, the adoption of feed-in tariffs and similar pricing schemes for renewable energy sources could make net zero energy a possibility for a much wider pool of homeowners.

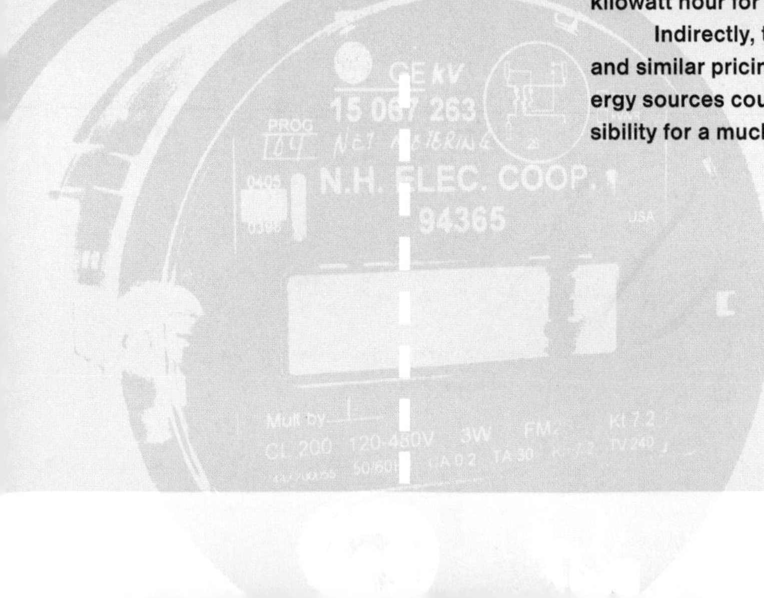

GRID-TIED PHOTOVOLTAIC SYSTEMS

In a grid-tied photovoltaic system, excess electricity produced by solar panels is sold to the local utility in an arrangement called net metering. When panels can't supply all of the power that's needed, electricity flows the other way. An inverter changes the DC power produced by the panels into AC used by appliances and other devices.

Without Battery Backup

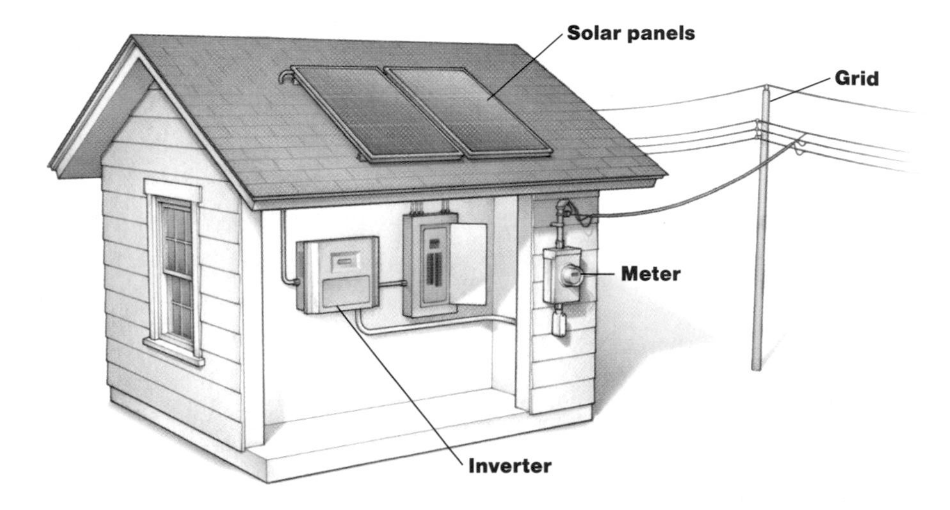

With Battery Backup

Grid-tied systems with a battery backup ensure that electricity will be available regardless of weather conditions or status of the utility's grid. Battery systems are more complex and more expensive.

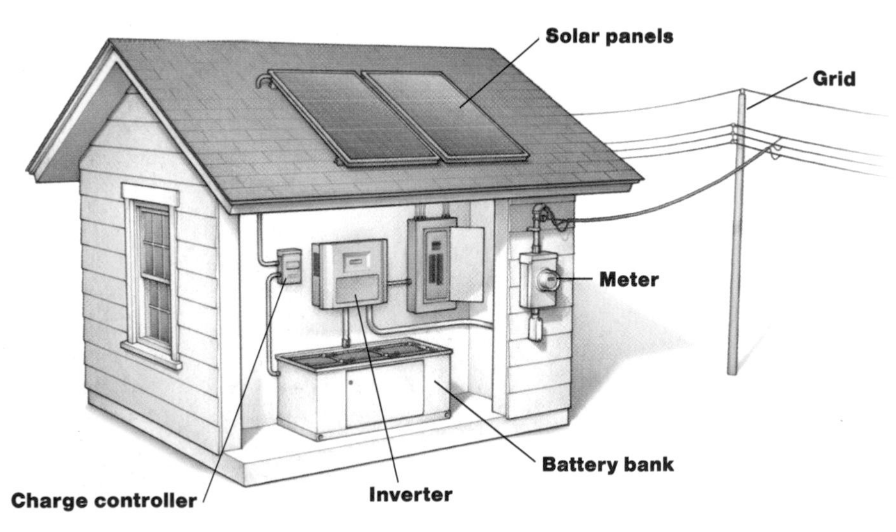

Adapted from U.S. Department of Energy

(continued from p. 106)

as well as building in maintenance costs that a grid-tied system doesn't have. More to the point, it's unlikely that an off-the-grid system can store enough electrical energy in batteries to reach net zero operation. Grid-tied systems that incorporate batteries may be the best of both worlds in areas where power outages are common and uninterrupted electricity is a high priority.

There are several kinds of photovoltaic cells, but by far the most common is made with thin wafers of silicon, sliced either from a single crystal or a cast block of crystals. The most efficient are those made from a single crystal (monocrystalline); wafers cut from multi-crystalline silicon are somewhat less efficient but also a little less expensive. Manufacturing solar cells from crystalline silicon wafers is relatively expensive, and it requires a high-grade of silicon.

Conventional crystalline cells are linked together in a weatherproof frame, typically aluminum, and protected from damage by a glass cover. This forms a module, or panel. A number of panels are grouped together into an array that can be mounted on top of a roof or in fixed or moveable racks on the ground.

SOLAR EFFICIENCIES CONTINUE TO RISE

Crystalline silicon commands more than 90 percent of the market. Cells that are packaged into residential PV systems operate at efficiencies of up to about 20 percent, about half of what's been accomplished on a small scale in the lab and significantly less than the cells used to power spacecraft. So there's still a lot of potential.

Solar electric panels are typically mounted on the roof, but they also can be arranged in racks on the ground when the roof isn't large enough, is incorrectly oriented toward the sun, or is too shaded.

Manufacturers also have found ways to make thin, flexible PV films by depositing amorphous silicon and other materials directly onto glass, stainless steel, or other materials. Very thin ribbons of silicon can be made in a different manufacturing process. Commercially available thin-film PV doesn't operate as efficiently as crystalline cells (up to about 11 percent efficiency), but it's less expensive to manufacture and can be deployed in more creative, less obtrusive ways.

Researchers also are learning how to use materials other than silicon to make photovoltaic cells, including cadmium telluride, copper indium gallium diselenide, and even plastic. In the short run, crystalline silicon is likely to remain dominant, but over time research will bring more efficient and less expensive forms of PV to market, not only helping the energy-efficient building boom in the United States but also offering real hope to the billions of people worldwide who currently don't have reliable electrical service of any kind.

No matter how the electricity is actually produced, there are several other components in a PV system collectively called the *balance of system.* Chief among them is the *inverter,* which changes the direct current (DC) produced by the panels into the alternating current (AC) that household appliances and lights run on. Off-the-grid systems need additional equipment, including a charge controller, which regulates how much power is used to recharge the batteries, the deep-cycle batteries, and

Photovoltaic panels are often made from silicon crystals that are sawn into thin wafers. But the material also can be formed into a continuous ribbon, such as this one manufactured by United Solar.

The direct current produced in photovoltaic panels must be converted to alternating current before it can power household appliances and lights. Inverters do the trick, although they consume a small amount of energy in the process.

possibly some kind of backup power like a generator. All systems have a DC disconnect that can separate electricity-producing components from either the grid or the rest of the house. Most systems also come with a meter that lets homeowners monitor energy production.

Racking up a photovoltaic system

The easiest way to assemble a residential PV system is to mount panels in fixed racks that are attached to a south-facing roof. For the best output, racks should be angled so they approximate the latitude of the site. If you live in Jacksonville, Florida, for example, at latitude 30 degrees north, the racks should be pitched at 30 degrees. In Philadelphia, 40 degrees north latitude, racks would be set at roughly 40 degrees (the equivalent of a 7-in-12 and a 10-in-12 roof pitch, respectively).

PV arrays also can be arranged to take advantage of local weather conditions (cloudy mornings, bright afternoons) or to produce more power at a particular time of day. West-facing arrays, for example, can be big afternoon producers of electricity and of interest to utilities that must meet their biggest peak loads at that time of day.

Roof-mounted racks have a few potential drawbacks. The roof might be pitched at the wrong angle, for example, forcing the builder either to settle for less-than-optimum performance from the solar panels or to jack the panels up to the correct angle and make them much more visible.

Grid-tied PV systems can have a battery backup, ensuring a supply of electricity even when the utility grid is down and the panels are unable to produce electricity. This may be the preferred approach in areas where utility service is spotty and prone to blackouts.

Panels that lay flat on the roof are the most unobtrusive visually, but when the angle of the roof is too shallow, racks can be elevated.

Some installers screw racks for solar panels directly to the roof and seal the connection with caulk. In addition to being more susceptible to a leak, this approach also means both racks and solar panels must be removed before the roofing can be replaced.

Many installers fasten the racks directly to the roof by driving lag screws through roof shingles and sheathing and into rafters or trusses. A glob of caulk is supposed to head off any potential leaks. A better approach is to use standoffs to support the racks. They're attached to the roof deck and framing before the shingles or other roofing materials go on. Water is kept out by a weatherproof boot, just like a plumbing vent. Standoffs make it a lot easier to replace roofing without a wholesale removal of racks and panels, and weathertightness doesn't rely on caulk.

Racks for PV panels can be fixed or adjustable. The most complex designs track the path of the sun across the sky, which boosts efficiency but also adds to installation and maintenance costs.

An advantage of using standoffs (top left) is that they do not have to be removed when the house is reroofed (bottom left).

Solar panels also can be installed on the ground if there isn't enough room on the roof or the roof is shaded by trees or in an otherwise less-than-ideal location. Panels set in fixed racks capture the most potential sunlight only during a limited portion of the day. To improve output, panels can be set in racks that track the motion of the sun, either manually or with an automatic motor drive. One- and two-axis frames are available.

Single-axis tracking is set to the appropriate angle of the sun above the horizon. The panels need to be reset each month or so. Dual-axis tracking follows the diurnal sun angle as well as its altitude from season to season. An automatic tracker can increase an array's output significantly, but the more complicated the system, the higher the initial cost and the greater the potential maintenance costs.

It's hard to make roof-mounted panels unobtrusive, especially if they have to be tilted away from the plane of the roof deck to catch the sun. Local ordinances, in fact, may prohibit rooftop installations, as was the case with a project in the downtown historic district in Santa Barbara, California (see "Bringing 1887 into the 21st Century" on p. 88). Manufacturers now make a variety of photovoltaic products that look just like roof shingles or that can be applied as a flexible membrane to a metal roof

Sunslates™, made by Atlantis Energy Systems, are one kind of building-integrated photovoltaics. The crystalline-silicon tiles take the place of regular roofing and blend into the background visually.

Roof integrated modules from General Electric® are just under 6 ft. long and 18 in. wide with upper and lower edges that overlap. They are made to be compatible with tile roofing.

Thin-film laminates, applied directly to metal roofing, virtually disappear. The 9-ft.-long PV laminates on the roof of a house in Newburyport, Massachusetts (above), are hard to spot unless you know they are there. At a home in Windham, New Hampshire (left), 19-ft.-long Uni-Solar® laminates cover a total of 1,200 sq. ft. of Englert standing-seam roofing, part of a 6kW PV system.

Manufacturers also offer photovoltaic cells incorporated directly into window glazing, allowing windows and skylights to serve a dual purpose, as in this semi-transparent amorphous silicon from BP Solar.

HIGHER TEMPERATURES, LOWER OUTPUT

The efficiency of solar electric panels is measured at the standard test condition of 25°C (77°F), but the nominal efficiency of a cell declines by between 0.4 percent and 0.5 percent for every degree Celsius above that. On one of those blistering summer days, high roof temperatures could easily depress electric output by 15 percent.

or shingle or even buried in the glass of a skylight. Some use conventional silicon PV technology; others use thin-film technology.

Building-integrated photovoltaics, as this category of PV is called, are compatible even with old-school architecture, like a center-chimney colonial. But most have an Achilles heel: heat. Unlike rack-mounted panels, building-integrated cells don't get air circulation around them so they're more likely to lose efficiency as the temperature goes up.

Racks of solar panels are not always a good fit architecturally, but building-integrated products minimize the visual impact—even on classic colonial houses.

Solar potential varies by region

Weather patterns and latitude affect the solar potential at any given house site, which is often expressed as kilowatt hours per square meter per day (kWh/m^2/day). Parts of Arizona, Nevada, Texas, New Mexico, and California have the highest solar potential in the United States; the Pacific Northwest, Michigan, and parts of the Northeast have the lowest. According to the National Renewable Energy Laboratory (NREL), the potential ranges from about 4kWh/m^2/day to about 7kWh/m^2/day.

Given the solar potential at a particular spot, a designer can predict how much electricity a PV installation will produce in a year. In sunny San Diego, a 4kW system should produce slightly less than 6,000kWh. Across the country in Concord, New Hampshire, the same system would produce about 17 percent less power, or roughly 5,000kWh. An online calculator at NREL's website (see Resources on p. 244) called PV Watts estimates electrical output based on the location and the capacity of the system. The software includes data for

SOLAR POTENTIAL IN THE UNITED STATES

Solar potential varies widely around the United States. Parts of the West have the most, whereas the Pacific Northwest, New England, and parts of the upper Midwest have the least.

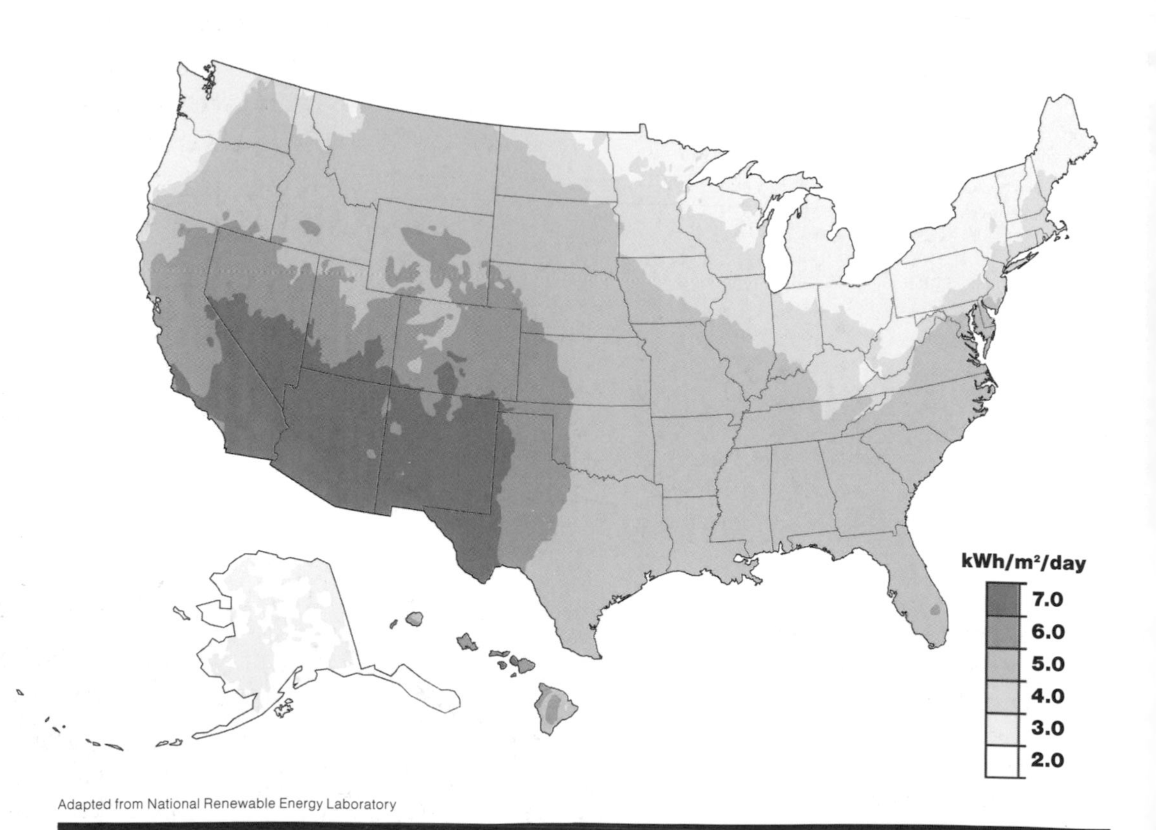

Adapted from National Renewable Energy Laboratory

hundreds of U.S. cities. It also calculates the dollar value of the power based on local electric rates.

If you're not living in a net zero house at the moment but you're curious about the practical meaning of all these numbers, gather your electric bills from the past year and add up the total (the average U.S. household uses more than 11,000kWh of electricity per year). You can then use the PV Watts program to quickly figure out the required capacity of a PV system that would meet 100 percent of your annual demand.

Keep in mind that a huge variable is the total electrical load. A key goal in net zero design is reducing the load to the point at which a practically sized renewable energy system can meet all the demand. At a Vermont zero energy house designed by architect David Pill, for example, the design load was about half the U.S. average (see "The Good Life in Vermont" on p. 138); at a Habitat for Humanity net zero project in Wheat Ridge, Colorado, it was halved again (see "Affordable Housing Meets Zero Energy" on p. 233).

This huge photovoltaic station in Boulder City, Nevada, has peak production capacity of 10MW. The owner, Sempra Generation, is planning an expansion nearby that will be nearly five times as big.

Costs: Coming down, but still high

By 2015, the government's Solar America Initiative hopes to make PV-generated electricity competitive with conventional forms of utility power, and that may be closer than it appears. A 10MW photovoltaic project in Nevada owned by Sempra™ Generation has reportedly been able to produce electricity for 7.5 cents per kilowatt hour using thin-film cadmium telluride cells. First Solar®, the company that produced the panels, doesn't comment directly on per-watt installed costs but does say it is able to manufacture the cadmium telluride cells that are nearly 11 percent efficient for about 93 cents per watt.

Sempra's Nevada installation uses cadmium telluride in the solar cells rather than silicon. The manufacturer, First Solar, says panels are nearly 11 percent efficient and can be made for 93 cents per watt. This panel was built in the company's plant in Frankfurt, Germany.

In general, though, the cost of PV electricity remains substantially higher than utility power at current energy prices. At the end of 2008, solar electricity cost about 30 cents a kilowatt hour compared with an average U.S. residential electricity rate of about 11 cents per kilowatt hour.

These numbers are averages. In some regions of the country–in New England and Hawaii, for example–the cost of utility power is much higher, so the cost ratio of photovoltaic to utility prices drops to two to one or, in Hawaii's case, even to par. In regions where the cost of utility power is much lower, of course, the relative cost of photovoltaics becomes that much less attractive.

Those are useful comparisons, but PV systems aren't priced by the per-kilowatt hour cost of electricity, at least not directly. Instead, the benchmark is the cost per installed peak watt–the electrical output under the most favorable conditions. Residential systems, on average, cost between $8 and $12 per watt. So a system that at peak production can generate 3kW of electricity should cost between $24,000 and $36,000 installed (before any rebates or tax credits). And just how much power is that? Think of it in terms of 100W light bulbs. Our 3kW system would power 30 light bulbs at the same time, which may not seem like much for that kind of money. PV's high cost explains why designers have to spend so much time reducing the electrical loads if they hope to get to a house to net zero. The power is too valuable to be wasted.

GETTING TO ZERO

State-by-state incentives for installing photovoltaic panels are all over the place, from nothing in some states to as much as $4 per watt elsewhere. New incentives passed by Congress after President Obama took office have changed the rules considerably and made renewable energy much more affordable. A state-by-state database of incentives and credits, called DSIRE, is available online (www.dsireusa.org).

Harnessing the Power of the Wind

Anyone who has driven across west Texas or southern Wyoming in recent years couldn't help but notice the hundreds of huge wind generators arranged neatly across the landscape. Texas is the U.S. leader in wind-generated electricity (four of the five largest wind farms in the country are located within its borders), but it is by no means the only state where utility-scale wind farms are under construction or already online. According to the American Wind Energy Association, total U.S. capacity by the end of 2008 was more than 21,000MW, enough to power the equivalent of 2.5 million homes and provide about 1.5 percent of the total U.S. electricity supply.

Wind generators on this scale are relatively new, and they've been getting a lot of public attention. But smaller, house-size wind turbines are attractive for many of the same reasons as photovoltaics, mainly predictable power costs made possible by a free, nonpolluting, and infinitely renewable natural resource.

Utility wind farms with scores of huge wind turbines are becoming an increasingly common sight on the U.S. landscape. The biggest turbines have rotor diameters of 300 ft. and produce 2MW of electricity each.

Huge blades manufactured at this Gamesa plant go to wind farms all over the United States. The multinational corporation has installed turbines that have produced more than 1,000MW of electricity since 2003.

Although the United States has enormous wind potential, challenges remain. In some areas, utility lines are inadequate to handle increased electricity production. Even getting the giant blades of utility-size turbines to the installation site can be a challenge.

Small wind generators can be part of a grid-tied or stand-alone system and can be paired with photovoltaics in a hybrid system. With a steady wind of the right speed (and barring mechanical problems), a wind turbine can produce electricity at a lower per-watt cost than photovoltaic panels and produce power around the clock. Wind turbines, however, have lots of moving parts, increasing the likelihood of repairs. Most need more room than an average suburban or urban lot can provide, and warranty periods are typically lower than those for photovoltaic panels.

Two basic turbine options, many tower options

The most familiar wind generators are two- or three-bladed turbines that sit atop a high tower and spin on a horizontal axis, like an airplane propeller. They come in a variety of sizes and electrical outputs.

Other versions rotate on a vertical axis. They take up less room, and some are designed to mount on the roof. That's an advantage for suburban or infill lots (Ed Begley Jr. and Jay Leno are two famous converts). There are some elegantly sculptural designs—such as the Savonius and Darrieus, both invented by European engineers in the early 20th century—but as extensive U.S. Department

WIND TURBINES NEED ELBOW ROOM

Horizontal-axis turbines are not for small lots. The National Renewable Energy Laboratory recommends a minimum lot size of 1 acre, and says the bottom of the rotor blades should be at least 30 ft. above any obstacle within 300 ft. of the tower.

Wind turbines sized for residential use are typically two- or three-bladed designs that sit atop a tower and turn on a horizontal axis. Tower height as well as site conditions are critical to how much electricity the turbine can produce.

of Energy (DOE) research proved in the late 1970s and early 1980s, vertical-axis turbines have lower aerodynamic efficiency and, at least anecdotally, have been prone to more mechanical problems than the best horizontal-axis designs. Rooftop turbines don't appear to be a very good bet despite their appeal.

Residential-size towers range from 60 ft. to 140 ft. Generally speaking, higher towers are better because they lift the rotors and generator into more turbulence-free wind and farther away from ground affects like trees, buildings, and rough terrain. Increasing tower height from 60 ft. to 100 ft., for example, can

A tilt-down tower allows a wind turbine to be lowered to the ground for periodic maintenance. A winch connected to a boom, or gin pole, at the base of the tower makes the operation possible.

BUILDING-INTEGRATED WIND TURBINES

Installing a wind turbine on top of the house where the electricity will actually be consumed is an appealing idea. But building-integrated turbines, as they are called, seem to suffer a variety of problems. According to an analysis by Alex Wilson published in *Environmental Building News*, potential safety hazards, vibration and structural stress, poor performance, and wind turbulence all are hurdles this budding industry has yet to overcome.

Emblematic of these difficulties is an experiment Wilson cites that was conducted in Wisconsin by Madison Gas and Electric. The company mounted a vertical axis turbine on a pole to mimic an installation on top of a single-story building and then measured how much power it produced. Although the turbine was rated at 10kW, it was never able to produce more than 600W of electricity "even on a very windy day." In fact, Wilson says, a real-time meter available online found that the turbine had produced only 33kWh of electricity over a stretch of four months.

Rooftop installations generally had a lot of trouble meeting rated capacities, Wilson found, leaving the very large utility-size turbines by far the best wind producers and the small tower-mounted residential turbines a very distant second.

boost power output by more than 25 percent, providing a homeowner can get around local height restrictions and deal with possible complaints from neighbors.

Towers fall into two basic categories, those with guy wires and those without. Self-supporting towers take up less room and aren't quite as visible, but they're also more expensive. A 120-ft.-tall self-supporting tubular tower, for example, may cost more than $30,000, about twice as much as a standard guyed-lattice tower of the same height. The problem with tall towers, of course, is that they put the turbine farther out of reach and make maintenance that much more difficult. For that reason, tilt-down designs are available.

The turbine, of course, is extra. Overall, Bergey Windpower Company, a manufacturer, estimates the cost of a 10kW turbine at between $48,000 and $65,000 installed. Roughly $40,000 of the total is in equipment, the rest in shipping and installation. This would make the cost per installed watt for wind a good deal lower than it is for photovoltaics, but Jim Green of NREL says the cost of wind-generated electricity depends heavily on the size of the turbine. At a capacity of 1kW, the

Under the right conditions, a residential wind turbine, like this one from Bergey, can produce electricity at a lower per-watt cost than photovoltaic panels. This guyed lattice tower is among the least expensive tower options.

Easy access to the turbine with a tilt-down tower makes maintenance a lot easier. This ARE-110 turbine is being installed at Appalachian State University.

per-watt costs are about the same. As turbine size increases, so do the economies of scale; with a 10kW turbine, Green expects a per-watt cost of between $6 and $8, slightly lower than PV but not dramatically so. Tower cost is another big variable, he says.

Electrical output changes dramatically with wind speed

Wind turbines are sized by their nominal output at a given wind speed. For example, Bergey's 10kW turbine (which was used on the house featured in "The Good Life in Vermont" on p. 138) should generate 10,000W when the wind speed is 31 mph, according to the manufacturer, but less power at lower wind speeds and more power when the wind speed is higher. A turbine made by Southwest Windpower called the Skystream 3.7® reaches peak power production of 2.4kW at a wind speed at 29 mph.

These relationships between wind speed and electrical output are called *power curves.* Manufacturers provide them for the turbines they make, although there is no guarantee they are completely accurate. (NREL also publishes reliable power curves for some turbines on its website.) There is no standardization in the industry for creating these values. One manufacturer's turbine may reach its nominal electrical output at a different wind speed than a turbine made by someone else.

MORE WIND, MORE ELECTRICAL OUTPUT

This power curve for a Skystream 3.7 turbine shows peak output at 29 mph. Small changes in wind speed have a big impact on electricity production.

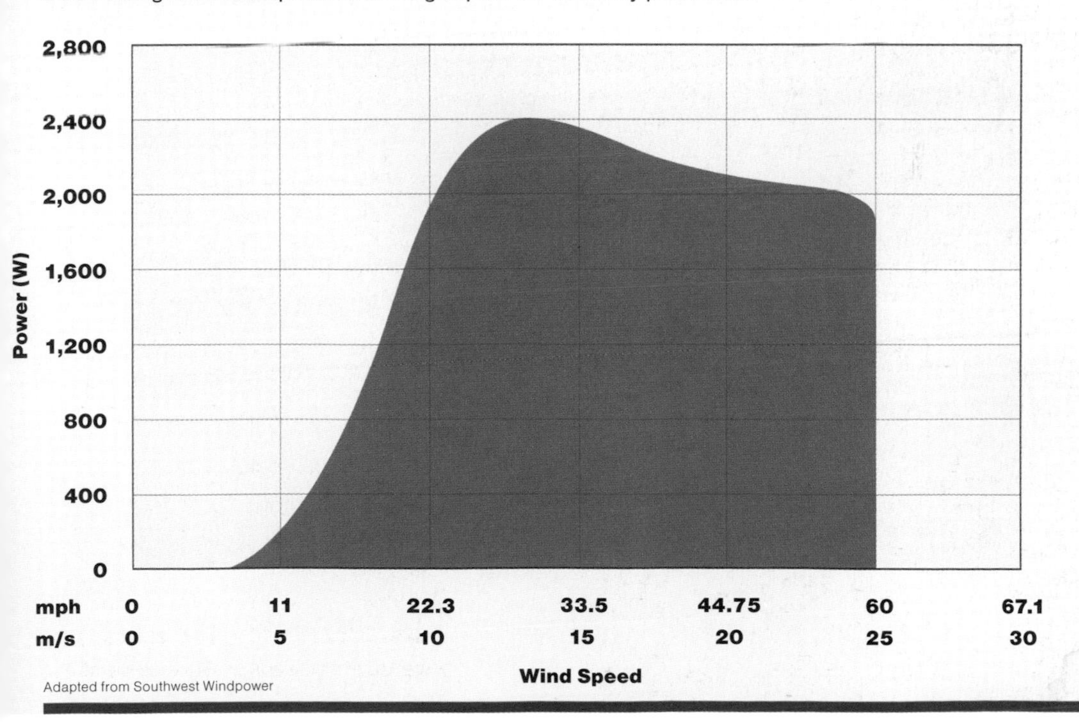

Turbines typically begin turning when wind blows 4 meters per second or so (about 9 mph), according to the American Wind Energy Association. Most turbines will disable themselves automatically at a given wind speed so they don't self-destruct.

Wind speed plays a huge role in how much power a turbine can produce. The relationship is cubic, meaning there's an eightfold increase in power when wind speed doubles. (Here's the math: Suppose wind speed increases from 10 mph to 20 mph; $10^3 = 1{,}000$; $20^3 = 8{,}000$; $8{,}000 \div 1{,}000 = 8$.)

All of this makes accurate predictions about annual electricity production somewhat difficult. For solar electric panels, free software will pop out an estimate of annual energy production in just a few seconds. There's no such program available for wind energy. For a rough estimation of a turbine's annual energy output in kilowatt hours per year, a homeowner or designer can multiply some simple values by a numerical constant. Here's the formula:

$$AEO = 0.013828 \times D^2 \times V^3$$

where AEO is the annual energy output, D is the diameter of the rotor in feet, and V is the annual average wind speed in miles per hour.

Let's take a hypothetical turbine with a rotor diameter of 20 ft. and an average wind speed at the house site of 10 mph. Using this formula, the turbine would generate about 5,500kWh of electricity annually ($0.013828 \times 20^2 \times 10^3$). With the wind speed only 1 mph more, annual output jumps to 7,300kWh a year, an example of how powerful the cubed rule really is.

Also useful for ballpark estimates of annual output is a turbine's *capacity factor.* That's the actual electrical output as a percentage of what it would be if the turbine operated at 100 percent of its capacity 100 percent of the time. For small-scale residential turbines, they typically operate in the 10 percent to 25 percent range. For an estimate of how much electricity a turbine would make in a year, multiply the rated power by the number of hours in a year (8,760) by the capacity factor.

A more accurate prediction comes from something called Wind CAD, which considers the turbine's power curve, average wind speed, and variables such as elevation and site turbulence. Similar calculators are available else-where, including one at the website of the Danish Wind Turbine Manufacturers Association.

Is there going to be enough wind?

Like solar potential, the nation's wind resource varies by region. Coastal New England, Texas, the mountains of Colorado, and the upper Midwest all have a lot of potential for wind generation, whereas parts of the southeastern United States have very little. What's different about wind potential is the affect that local topographical features have on a site. A wind generator might be feasible on one side of town and a waste of money on the other.

A map produced by NREL shows wind potential across the country, but it's only a place to start. It may take careful monitoring over an extended period of time to determine whether site conditions are right for a wind turbine. Annual wind speeds would have to average between 9.8 mph and 11.5 mph—Class 2 on the U.S. Department of Energy's scale—for a wind generator to make economic sense, according to the Wind Association.

Wind velocities are often higher in winter than in summer, making a good case for hybrid systems that combine photovoltaics and wind turbines. Some commercially available systems, in fact, are sold as wind/PV packages.

NREL estimates that wind generators could meet at least 20 percent of U.S. electrical needs by 2030 (Denmark has already reached this point). On the plus side, it can be cheaper to make electricity with a wind generator than with photovoltaics, and wind generators produce power night and day, rain or shine. The problem is the inherent maintenance and repairs that go along with small-scale wind turbines. Whereas warranties on PV panels often run to 25 years (solid-state components, no moving parts), turbine manufacturers may limit guarantees to 5 years. Even long-time wind advocate Paul Gipe would take photovoltaics over a house-size wind

WIND POTENTIAL IN THE UNITED STATES

Wind turbines don't become economically attractive until average wind speeds hit about 9 mph. Some parts of the country have more wind potential than others.

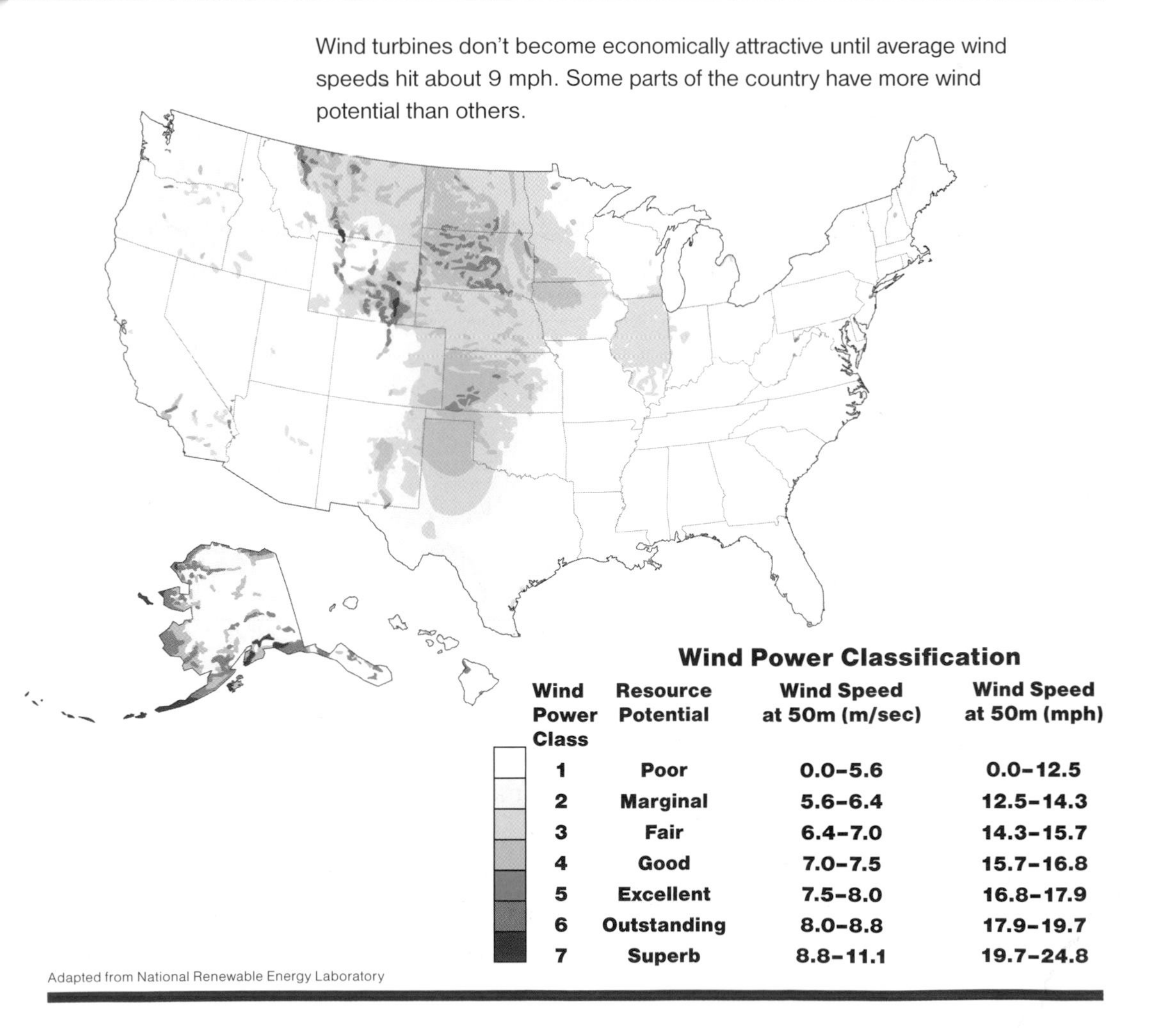

Wind Power Classification

Wind Power Class	Resource Potential	Wind Speed at 50m (m/sec)	Wind Speed at 50m (mph)
1	Poor	0.0–5.6	0.0–12.5
2	Marginal	5.6–6.4	12.5–14.3
3	Fair	6.4–7.0	14.3–15.7
4	Good	7.0–7.5	15.7–16.8
5	Excellent	7.5–8.0	16.8–17.9
6	Outstanding	8.0–8.8	17.9–19.7
7	Superb	8.8–11.1	19.7–24.8

Adapted from National Renewable Energy Laboratory

turbine. Why? Better reliability. Even so, some experts feel more reliable turbines are finding their way into the market.

Making Hot Water with the Sun

Solar hot-water systems are a natural fit with net zero houses. They're capable of meeting much of the demand for domestic hot water, and in some instances can help with space heating as well. Hot water accounts for a significant amount of energy use for most U.S. homes—about 17 percent overall, according to the government's energy office—so letting the sun do the work means lower energy bills and lower carbon emissions.

There are a variety of solar water heaters on the market, some suited only for warm climates and some designed specifically to cope with subfreezing temperatures. No matter where you live, there's a solar collector out there that will work.

One easy way of dividing the field is to think of collectors as either active or passive systems. Active collectors use electric pumps to move a fluid (either water or antifreeze) through the collectors and into a storage tank. Passive systems, which are simpler and less expensive, rely on gravity and convection to move water through the collector. There are no electrical components to break, and so these systems tend to be less expensive and more reliable.

One disadvantage of passive systems is that they're more susceptible to freezing pipes. Researchers are looking at a variety of options to make them more reliable in cold weather areas and thus more available to a greater number of U.S. households.

MAKING PASSIVE SOLAR WORK WHERE IT'S COLD

Water lines that freeze and burst have been one factor limiting the wider use of passive solar hot-water systems. The U.S. Department of Energy (DOE) reports that the current practice of using insulated copper pipes limits the use of these systems to the southern tier of the country and makes truly risk-free installation possible only in Florida and along the southern California coast.

One promising avenue of research involves pipes that can freeze and thaw repeatedly without damage along with the use of freeze-protection valves that protect water lines by circulating a small amount of warm water when the temperature dips toward freezing.

The National Renewable Energy Laboratory conducted tests on several brands of pipe made from cross-linked polyethylene (PEX). Some of them were able to withstand 400 freeze–thaw cycles without breaking; one brand could tolerate only 10 cycles. One problem with PEX is that it has an upper temperature limit of 210°F, the DOE said.

Freeze-protection valves consume a certain amount of water. But if 1,000 gal. per year is deemed acceptable, and the pipe issue could be solved, the department concluded the use of passive systems could be much wider.

Some of the basic types of collector are as follows:

- **Flat-plate collectors.** A basic flat-plate collector, which can be up to 4 ft. wide and 12 ft. long, consists of a shallow, insulated housing that encloses tubes for the water or heat transfer medium and a plate that helps absorb solar energy. A sheet of glass covers the top of the collector. Fluid flows through the pipes, picks up heat from the sun, and is moved into a hot-water storage tank or directly into the domestic hot-water supply. In an integrated storage collector, potable water is heated and stored in tubes or tanks inside the collector.

FLAT-PLATE COLLECTOR IS STANDARD

A flat-plate collector, the most common type of hot-water collector, consists of a series of tubes inside a glass-covered frame. Water or a heat transfer fluid picks up energy from the sun to heat the domestic water supply.

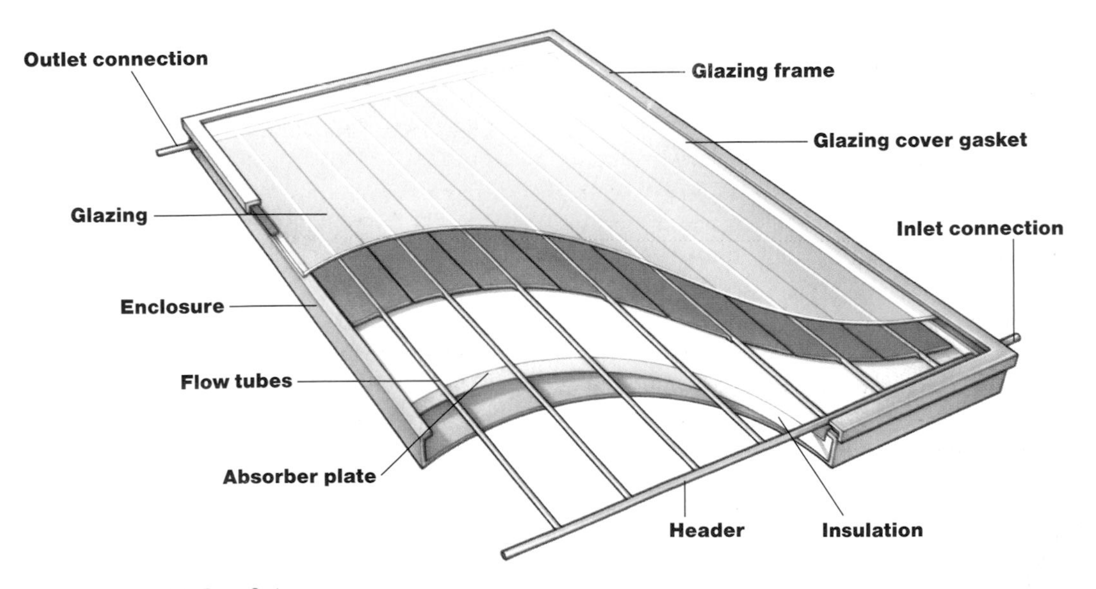

Adapted from Florida Solar Energy Center

- **Evacuated tube collectors.** These collectors consist of parallel rows of glass vacuum tubes. Inside, a fin absorbs heat from the sun and transfers it to a liquid medium. Evacuated tube collectors are sometimes specified in cold climates because of their high efficiency. They are more expensive than conventional flat-plate collectors.

- **Thermosiphon collectors.** Natural convection moves water through the collector without the need for pumps. As warm water rises into a storage tank, cooler water flows in the bottom to replace it. Some versions can be used in frost-prone areas, but these are typically found in climates that don't have freezing temperatures.

EVACUATED TUBE COLLECTOR

An evacuated tube collector consists of pairs of glass tubes separated by an evacuated space (a vacuum). A heat transfer fluid in the inner tube absorbs solar energy to heat water.

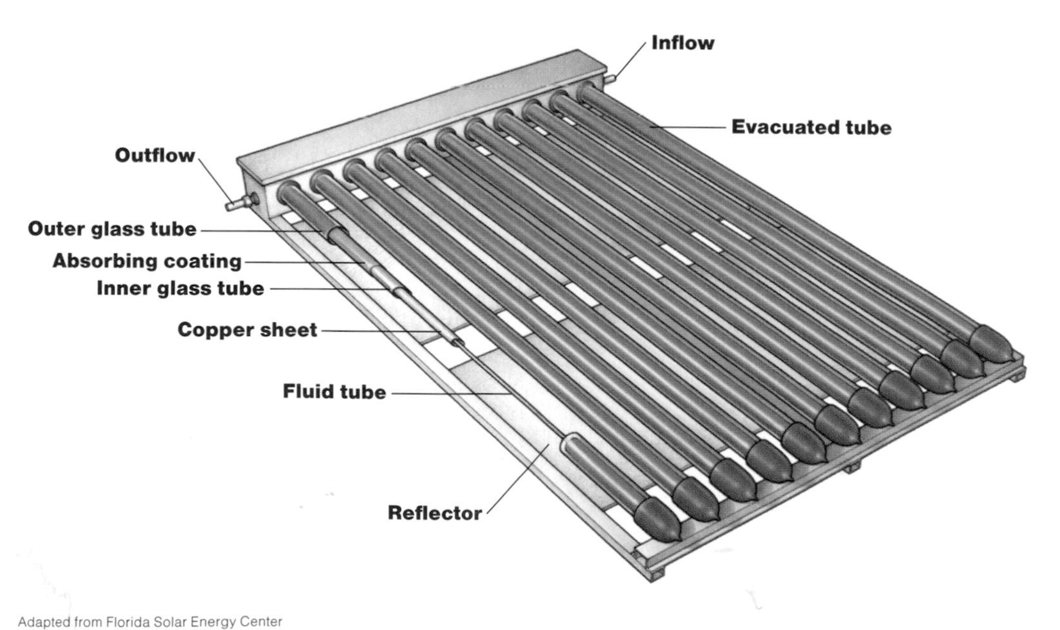

Adapted from Florida Solar Energy Center

CIRCULATION BY NATURAL CONVECTION

Thermosiphon solar collectors do not rely on mechanical pumps to circulate water. Instead, they rely on natural convection to move water from the collector to a storage tank, which is located above the collector.

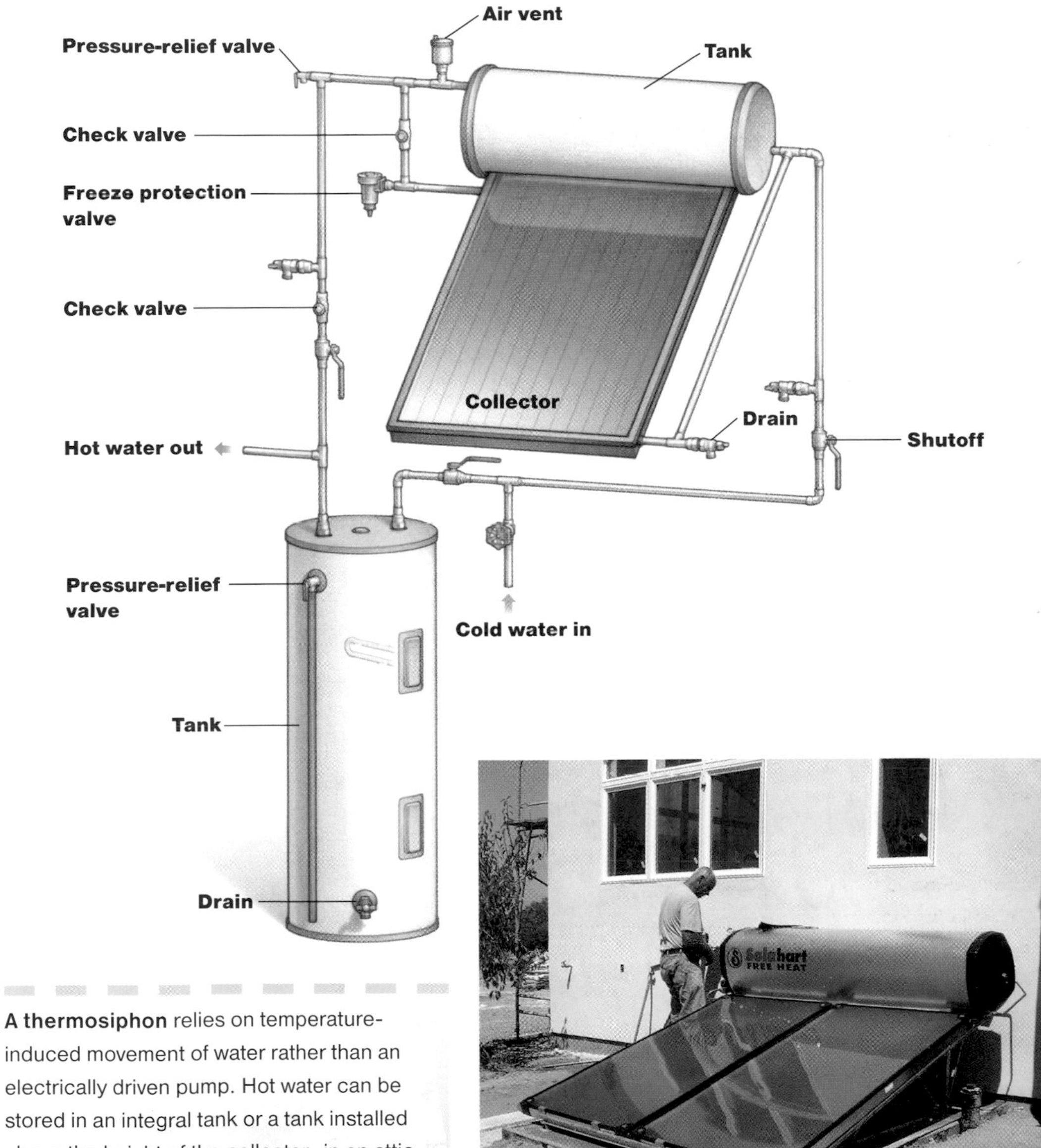

A thermosiphon relies on temperature-induced movement of water rather than an electrically driven pump. Hot water can be stored in an integral tank or a tank installed above the height of the collector—in an attic, for example.

Adapted from Florida Solar Energy Center

DRAIN-BACK SYSTEM FOR COLD CLIMATES

In areas where freezing temperatures threaten solar collectors, a drain-back system is a good option. When the system is not actively collecting heat, water from the collector and outdoor piping drains into an insulated tank. When the sun is shining and the system resumes operation, water is returned to the collector with a pump.

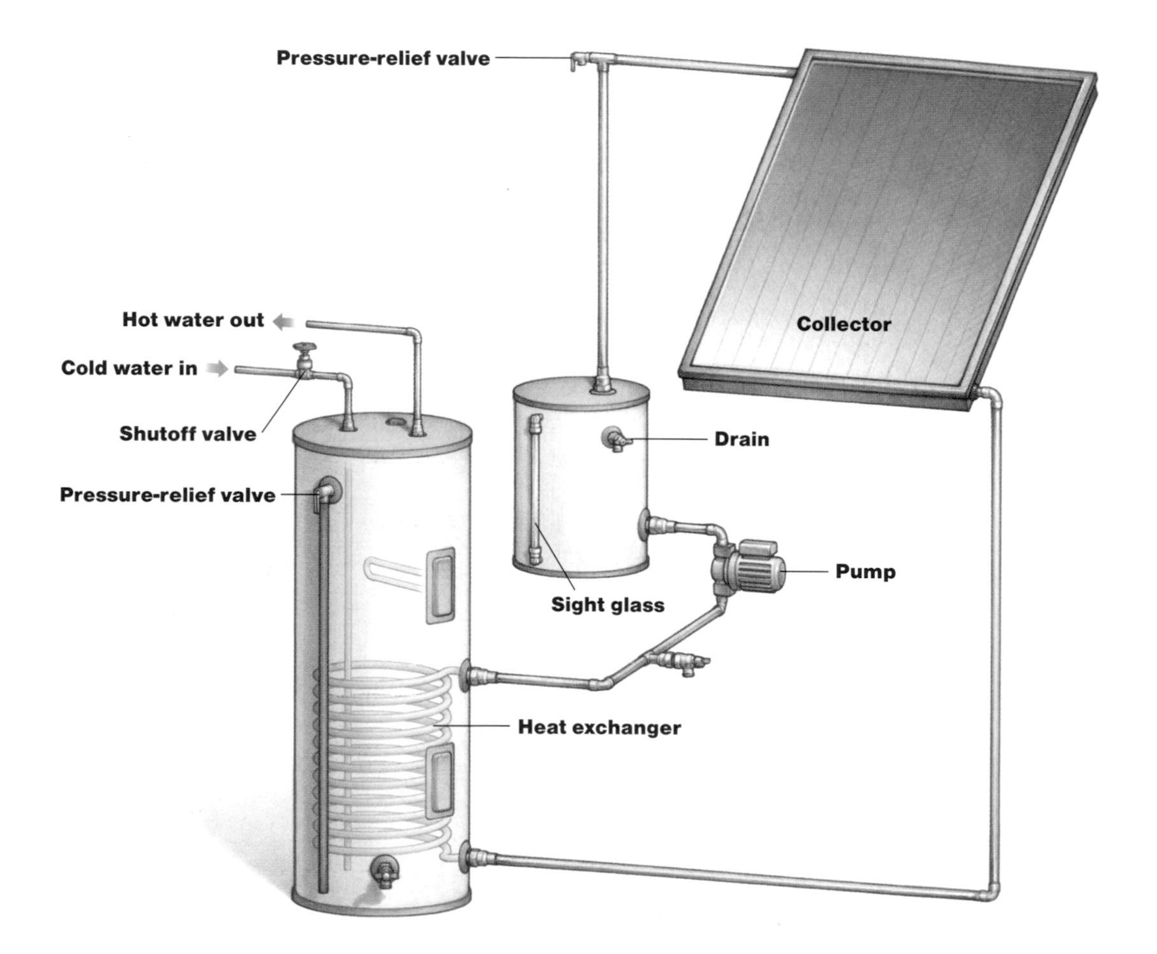

Adapted from Florida Solar Energy Center

DIRECT PUMPED SOLAR COLLECTOR

In a typical direct pumped system, water is heated in the collector and pumped to a storage tank, which supplies household hot-water needs. An indirect system is similar, but heat is transferred from the collector to the hot-water tank via antifreeze and a heat exchanger in the storage tank. These systems are appropriate for cold parts of the country.

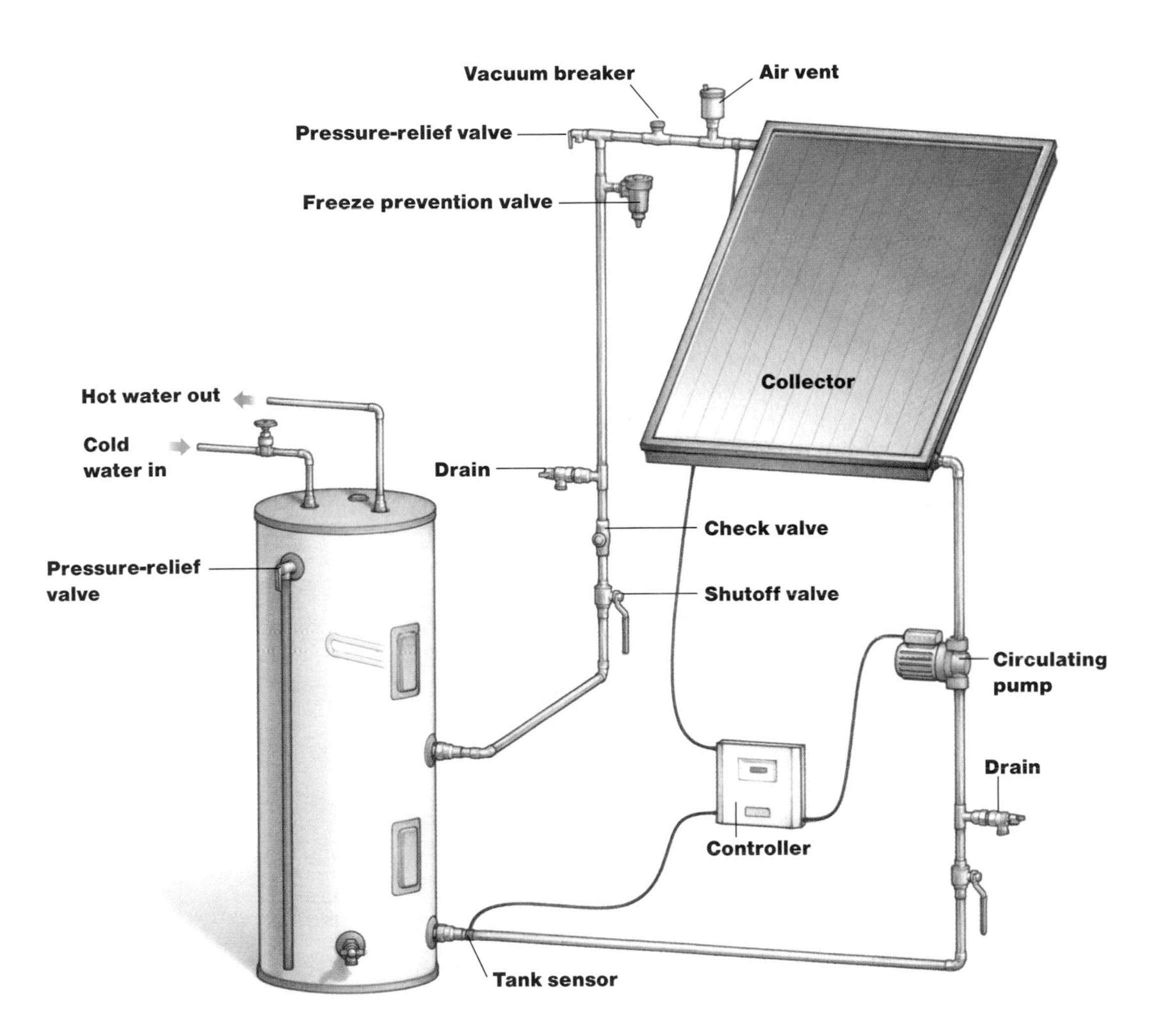

Adapted from Florida Solar Energy Center

There are many variations on these basic themes. Drain-back systems, for example, remove all water from the collector and exposed piping when the system isn't collecting heat as a hedge against freezing. Water drains back into an insulated reservoir and is pumped back into the collector when the system resumes operation. Hot-water systems can be combined with a small photovoltaic panel that operates the pump.

In most systems, water goes from the collector to a storage tank where the temperature can be maintained by a supplemental source of heat—electricity, natural gas, fuel oil, or liquefied petroleum. Systems are often sized on the assumption that each person in the house will use 20 gal. of hot water per day. Consumption, however, can vary quite a bit, depending on personal habits and family size.

Solar hot-water systems typically include a storage tank (above left). Water may need a boost from a conventional fuel source before it's piped to sink or shower for use. These tanks are connected to a pair of solar collectors on the roof (above right). If water in the system gets too hot, it's piped to a radiator in an adjoining basement area to cool down.

Solar collectors also can heat water for space heating, but there are some caveats. Because sunlight is weaker just at the time of year when heating loads are the highest, there are limits to how much of a contribution solar collectors can make. Moreover, some heating systems are better suited to a solar boost than others. A radiant-floor system, for example, may need water at only 100°F on most days, whereas a conventional hydronic heating systems needs water at 160°F or more. Homeowners in conventionally built houses probably shouldn't expect to get more than about one third of their winter heat from solar, although a superinsulated house with radiant-floor heat should be able to do better than that.

An obvious drawback to solar space heating is that a call for heat is likely to come right around the time the sun is going down and temperatures are dropping. In other words, just as the solar collectors are knocking off for the day they're needed the most. Tanks that can hold hundreds or even thousands of gallons of water can help by storing heat generated during the day. When collectors are no longer picking up any energy, the heating system can tap into energy stockpiled during daylight hours.

INTEGRAL COLLECTOR

Integral collectors combine solar energy collection and water storage in a single unit. They are best suited to mild climates.

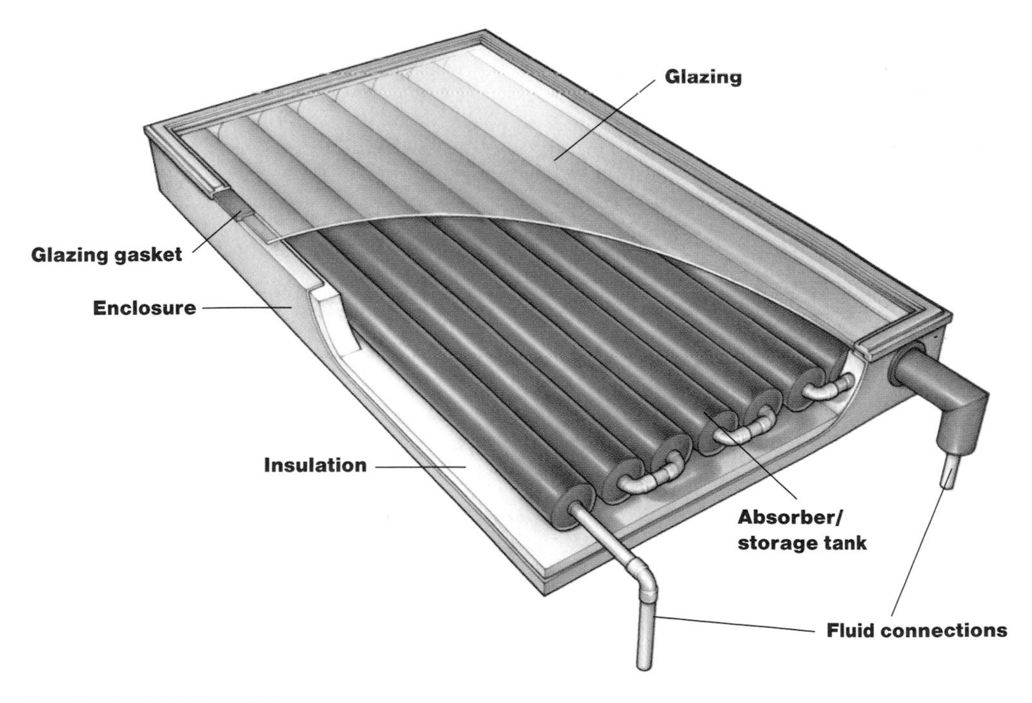

Adapted from Florida Solar Energy Center

Warm weather systems are the least expensive

There are lots of variables that affect the cost of a system. An installed system in Florida, where long spells of freezing weather aren't an issue, should cost between $3,500 and $5,500, according to the Florida Solar Energy Center. In Maine, an evacuated tube system for a family of four could top $10,000 before any incentives or rebates. When the cost of a solar hot-water system is rolled into a mortgage and amortized over the life of the loan, the added cost may be less than monthly savings in electricity or gas used for a conventional hot-water tank. In a net zero house, the added benefit of renewable energy taking the place of fossil fuels or electricity from the grid is obvious.

Federal and state incentives also lower the cost, sometimes substantially. Despite a depressed housing market, the National Association of Home Builders reported at the end of 2008 that increased federal incentives were helping Florida builders sell houses with both photovoltaic panels and solar hot-water systems at prices competitive with conventional houses in the same market.

Deciding when solar hot water makes sense

Solar hot-water collectors are common in net zero designs but not exactly a given. If the electrical output of other renewable energy systems (wind or PV) is sufficient, it probably makes more sense to heat water that way, avoiding the initial cost of solar collectors as well as any maintenance or repairs to the system. Savings can be sizable.

Deciding when to add more PV and when to invest in solar hot water can be a vexing question. What pays the best dividends may not always be obvious, and sometimes the unpredictable habits of whoever is living in the house can throw the best calculations right out the window.

Such was the case at the Habitat for Humanity house in Lenoir City, Tennessee, built under the guidance of the Oak Ridge National Laboratory (see "All Electric House: $1.16 a Day" on p. 210). Designers used computer modeling to look at solar hot-water options and settled on a system with 96 sq. ft. of collectors on the roof and a 200-gal. storage tank—expensive but apparently a good long-term bet given assumptions on how much water the family was likely to use.

But the family used 20 gal. a day of hot water when 63 gal. had been predicted. This disparity made designers wonder whether the $7,000 collector, plus the $1,300 tankless backup heater, was the best choice. It turns out that increasing the size of the PV system slightly and using an electric hot-water heater would have cost about $5,100 compared with the $9,300 total for the solar system that had been installed. "The only conclusion that can be drawn from this simple analysis is that the conventional wisdom may not be true in all cases and additional investigation into the comparison of solar water heating and PV investments may be warranted," an Oak Ridge report on the house concluded.

Sizing a Renewable Energy System

Once decisions have been made about the house—insulation, windows, solar orientation, mechanical systems, and the like—it's time to work in renewable energy. And here, there is no magic, just a methodical accounting of how much electricity will be consumed day by day over the course of a year. Energy modeling software (such as Energy-10 developed by NREL) can be a huge help.

Appliances and heating and cooling equipment are major energy consumers. Planners can check specifications provided by manufacturers to see how much electricity a particular appliance will use. Washing machines, dishwashers, and refrigerators all come with government-mandated labels showing predicted energy use. Likewise, heat pump manufacturers provide data on energy use, although planners may want to dive deeper than that and do more thorough calculations. Lighting fixtures have a known wattage.

Plug loads are a wild card, not only because phantom loads (the electricity surreptitiously used to power electronics and other devices when they are turned off) can be substantial but also because we keep finding more stuff to plug in.

In an all-electric house, a renewable energy system can be sized once these values have been determined. When other fuels are used, however, there's one more step: producing enough surplus electricity to compensate for the firewood, natural gas, liquefied petroleum, or other fuels that might be used for space heating or cooking.

U.S. Government — Federal law prohibits removal of this label before consumer purchase.

ENERGYGUIDE

Refrigerator-Freezer
- Automatic Defrost
- Side-Mounted Freezer
- Through-the-Door Ice

XYZ Corporation
Model ABC-L
Capacity: 23 Cubic Feet

Estimated Yearly Operating Cost

$67

$57 — $74

Cost Range of Similar Models

630 kWh

Estimated Yearly Electricity Use

Your cost will depend on your utility rates and use.

- Cost range based only on models of similar capacity with automatic defrost, side-mounted freezer, and through-the-door ice.
- Estimated operating cost based on a 2007 national average electricity cost of 10.65 cents per kWh.
- For more information, visit www.ftc.gov/appliances.

Government EnergyGuide labels are helpful in comparing the electrical consumption of one appliance to another. Choosing EnergyStar-qualified appliances is another way of reducing electrical consumption.

Conventional incandescent light bulbs turn most of the electricity they consume into heat, not light. Compact fluorescents produce more light for the same amount of power, so bulb wattages can be decreased, and they last much longer.

Homeowners who watch their use of electricity carefully, who remember to turn out lights when they're not needed, and unplug the TV instead of just turning it off, will get by with less renewable energy capacity than an energy spendthrift. But in the end, estimating the amount of electricity the house will use in a year is simply a matter of adding up the numbers, and taking a few educated guesses along the way. "You build a load from the bottom up," says Vermont energy consultant Andy Shapiro.

Look at All the Renewable Energy Options

Very active research on sources of renewable and nonpolluting energy is certain to open new doors in the future. For the moment, photovoltaics, wind energy, and solar hot-water collectors are the basic tools that designers and builders have at their disposal to get a house to zero energy performance.

Many houses will combine systems, using both solar hot-water collectors and photovoltaics or a wind generator. But the exact mix of energy resources depends greatly on the site, the house specifications, and the habits of its occupants. There is no single answer, just a single objective. The one constant is a super-efficient building envelope—well insulated and well sealed.

Deciding where, exactly, to spend the building budget can be a challenge. Paul Norton, a senior research engineer at NREL, says "*the* critical question" in net zero building is how much money to invest in renewable energy and how much to invest in improving the energy performance of the building itself.

At first, making energy improvements to the building is relatively inexpensive. Air sealing, for instance, doesn't require a huge cash outlay. When energy improvements move to the realm of triple-glazed windows and thick layers of sprayed-in polyurethane foam insulation, costs go up. Eventually, it's more expensive to make the building more efficient than it is to add renewable energy capacity.

Norton believes this crossover point occurs when 50 percent to 60 percent of total energy savings have been reached. That is, the first half of the journey toward zero energy takes place on the building side and the rest with renewable energy. As the cost of renewable energy continues to fall—either through more generous government subsidies or basic improvements to the technology—the equation will change as well.

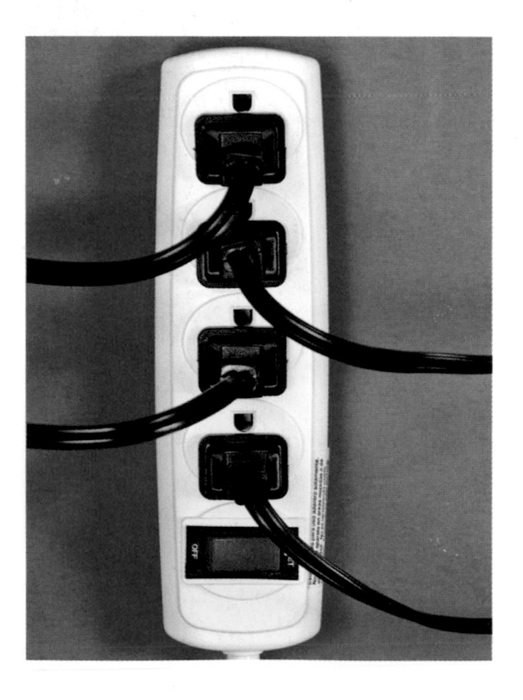

Plugging computers, televisions, and other electronic devices into a power strip instead of a wall receptacle allows power to be completely shut off with a single switch. Many devices plugged into the wall continue to draw a small amount of power even when they are turned off. The trickle of wasted energy adds up.

FUEL CELLS MAY POWER HOMES IN THE FUTURE

Fuel cells convert the chemical energy of fuel directly into electricity without combustion. When combined with photovoltaic or wind electrical generation, they become a flexible, renewable energy source that holds tantalizing possibilities for residential use.

Most current fuel cells use a fossil fuel, such as natural gas or liquefied petroleum, but fuel cells can also be run on oxygen and hydrogen. These cells throw off both electricity and heat with water as the only by-product; they're completely nonpolluting.

One key question is where the hydrogen comes from. Although it's an abundant element on earth, there are no deposits of gas we can tap into. Instead, it must be extracted from another material. On an industrial scale, hydrogen is typically made by mixing steam and methane in a process called *steam methane reforming.* But hydrogen can also be removed directly from water using an electric current in a process called *electrolysis.* When the electricity comes from a renewable source—photovoltaic panels, a wind generator, or a hydroelectric station—hydrogen becomes a way to store solar power.

In a project at Humboldt State University, for example, photovoltaic panels produce electricity that's used to run a compressor at a marine lab. Excess electricity is used to separate oxygen and hydrogen. Hydrogen is stored under pressure in tanks and used at night to run the compressor.

There are cost and technical challenges to this approach, and it will probably be some years before these interrelated systems are commercially available for residential use. But in time hydrogen may be widely used to heat and power homes.

Hydrogen fuel cells make electricity through an electrochemical process that combines hydrogen and oxygen, not by combustion.

Case Study

The Good Life in Vermont

Simple forms, sophisticated systems, and net zero energy consumption

David Pill and Hillary Maharam's home just south of Burlington, Vermont, is an appealing blend of simple architectural forms and unpretentious materials that fits right into its rural setting. More to the point, the Leadership in Energy and Environmental Design (LEED) Platinum building won the couple a $10,000 prize from the Northeast Sustainable Energy Association in 2009 for documented net zero energy performance.

SPECIFICATIONS

House size: 2,800 sq. ft.

Average heating degree days at site: 7,900

Wall and roof construction: Stick-built 2×6 and 2×10, respectively

Insulation type: Above-grade walls and roof, closed-cell spray foam full thickness of cavity; basement, 4 in. EPS rigid foam under slab; foundation walls, 2 in. EPS on inside of foundation walls plus 2×4 framed wall with blown-in cellulose; basement ceiling, 6 in. denim batt

R-values: R-16 under slab; R-21 basement walls; R-21 basement ceiling; R-40 walls above grade; R-58 roof

Windows: Thermotech triple-glazed fiberglass; U-0.15 on north, east, and west elevations; U-0.17 on south elevation

Wind/photovoltaic capacity: 10kW net-metered Bergey wind turbine

Heating source: 3-ton ground-source heat pump

Air-conditioning: None

Cost per square foot: $200

Designer: David Pill, Pill-Maharam Architects

Energy consultant: Andy Shapiro, Energy Balance

General contractor: Jim Huntington, New England Housewrights

A house designed by architect David Pill for his family in Vermont blends traditional architectural elements with state-of-the-art energy performance. Documented net zero performance won Pill a $10,000 prize from the Northeast Sustainable Energy Association.

Pill, an architect, designed the house for the 44-acre site after he and Hillary decided to leave the Boston suburbs with their two children for life in the country. Their goal was an affordable house with low environmental impact and no carbon emissions.

The result is an all-electric house, completed in 2007, that uses no fossil fuels and produces all the electricity it needs with a net-metered 10kW Bergey wind generator. It keeps everyone warm and comfortable on sunny winter days even with the heat turned off. Construction costs were about $200 per square foot—not exactly cheap, but less than what many custom homes cost these days, zero energy or not.

Nothing Exotic in Building Design

The house uses advanced framing with 2×6 studs 24-in. o.c. and 2×10 roof rafters. Pill eliminated all unnecessary framing members, not only to reduce waste but also to leave as much room as possible for insulation. He used two-stud corners, for example, and headers over doors and windows only where they were needed structurally (unlike houses framed with optimum value engineering, this one has structural sheathing).

Closed-cell polyurethane foam (R-6.5 per inch) completely fills the wall cavities of this trimmed-down frame, and 1 in. of polyisocyanurate foam wrapped around the outside reduces thermal bridging through the frame (see chapter 1). The 2×10 roof frame also is filled to capacity with sprayed-in foam. The basement slab is insulated with rigid foam.

Insulating the concrete slab of the basement is another step in reducing energy losses in winter.

A layer of rigid foam insulation on the outside of the house reduces thermal bridging between wood framing and cold outside air. Advanced framing techniques also helped by eliminating unnecessary framing material and leaving more room for insulation in the walls.

House shape and orientation both proved crucial to energy performance. The 60-ft. by 20-ft. house runs on an east-west axis, which creates a wide southern exposure to gather light and heat in the winter. Because the house is narrow, sunlight can reach every room and reduce heat and light requirements during the day. On the second floor, roof overhangs nearly 2 ft. wide block intense summer sun but let it in when the weather is colder.

On the first floor, a ground and polished 4-in.-thick concrete slab poured over a frame of wood I-joists soaks up the sun's energy on winter days and releases it gently at night. Windows are triple-glazed with insulated fiberglass frames made by Thermotech, a Canadian manufacturer. Windows have very low U-values and solar transmittance designed for their particular orientation—higher on southern exposures, lower on the rest of the house.

All of these features were fine-tuned by Pill's energy consultant, Andy Shapiro, who used Energy-10 software developed by the National Renewable Energy Laboratory's Center for Building and Thermal Systems.

Taking Advantage of Great Wind Potential

Pill and Maharam had decided early on that the house would consume no fossil fuels, and that made an all-electric design more or less a given. "The more we looked at it, the more we looked at the loads of the house, it looked

SUPER INSULATION YIELDS NET ZERO PERFORMANCE

David Pill's design for his family's Vermont net zero home includes a variety of building materials that sharply reduce energy consumption: polyisocyanurate sheathing, sprayed-in polyurethane foam, high-performance windows, and a thick layer of rigid foam insulation under the basement slab.

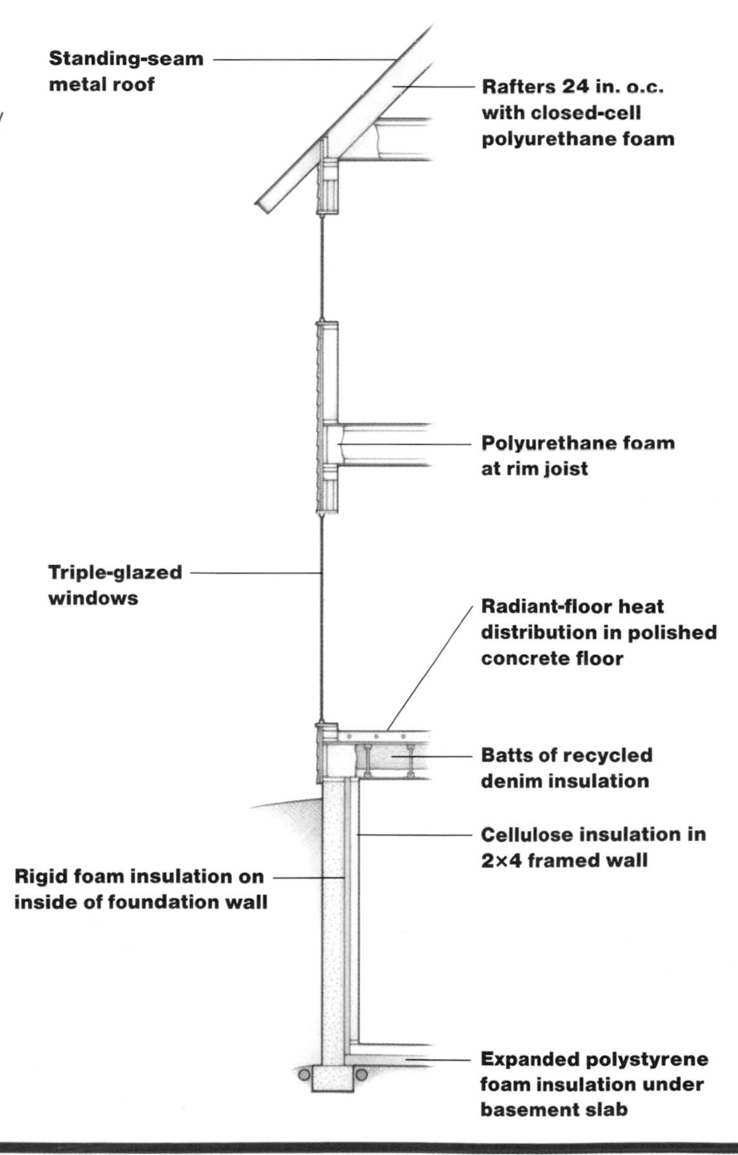

very possible to meet all our demands with a ground-source heat pump without using solar hot water," Pill says. "We wanted to stick to one simple system, either all [photovoltaic] or a wind turbine and consolidate the systems."

They caught a break on the site. Pill had noticed trees were "flagged," or distorted, from prevailing winds, possibly the result of unusual topography in the area. As it turned out, a neighbor worked for a company that makes wind assessment systems. He told Pill that wind on the site was obviously so good that it would be a waste of money to install monitors to verify suitability for a wind generator. A local turbine installer agreed.

The 10kW system cost $40,000 but it was also eligible for a $12,500 rebate from Vermont and a $3,000 federal tax credit, reducing its overall cost by nearly 40 percent.

From June 4, 2008, to June 4, 2009, the house used 5,999kWh of energy, plus the equivalent of 200kWh of firewood, while the generator produced 6,622kWh. With vigilance on the family's part, it was a net producer of electricity, by 423kWh.

The Pills' building site was breezy enough to make the wind turbine a good bet without extensive monitoring. On a lot with more marginal wind potential, testing is a prudent step before investing in a turbine.

A radiant-floor system provides heat. The thermal mass of the polished concrete slab is an energy bank that stores heat and helps even out sudden changes in indoor temperature.

A BALLPARK ESTIMATE FOR WIND-GENERATED POWER

Estimating how much electricity a small wind turbine will produce gets complicated. The amount of power varies dramatically with small ups and downs in wind speed, and wind isn't entirely predictable. Nor do turbines always produce as much power as manufacturers say they will. But there are ways to make a good guess.

Start with the "capacity factor," which is the actual electrical output as a percentage of what it would be if the turbine ran at 100 percent of its capacity 100 percent of the time. For small turbines, such as the one at this house, the capacity factor would typically be between 10 percent and 20 percent.

Now multiply the rated power of the turbine by the number of hours in a year (8,760) by the capacity factor. In this case, annual production should be at least 8,760kWh.

This turbine produced somewhat less than that, but even a minor bump in average wind speed on the site would change the result.

Geothermal Heat Shares a Deep Well

The house is heated by a radiant floor system connected to a 3-ton (36,000 Btu) ground-source heat pump. It draws water from the same well that supplies the family with drinking water. Water is normally returned to the well after it circulates through the heat pump. A variable frequency drive on the water pump can sense whether water is being drawn for domestic water (higher resistance) or the heat pump (lower resistance) and adjust flow accordingly.

The heat pump also produces domestic hot water, with an assist from a gray water heat-recovery system, a 5-ft. section of copper-wrapped waste line that picks up residual heat in shower water going down the drain. Pill says the equipment appears to be reducing the cost of heating water by about 30 percent.

Using the same well for domestic water and the heat pump reduced the cost of the heating system considerably, but it required a well that produced a lot of water. And here again, the couple got lucky. At 300 ft., the well was producing only about 2 gal. a minute—not enough, even, for domestic supplies.

Only the bathrooms on the second floor of the house were equipped with radiant-floor heat distribution. But the house's passive solar design has been enough to keep bathrooms warm enough without it.

They kept going, and at 480 ft., the well was producing 4 gal. a minute. Still not nearly enough.

A dowser had looked over the site and predicted that well drillers would tap into a vein producing at least 12 gal. a minute. "I was thinking, 'Oh, boy, now what?' " Pill says. "We were going to have to change our whole concept with the ground-source heat pump because 4 gal. a minute is not enough flow."

The dowser returned, and insisted the water was there. Well drillers had stopped drilling but left their pumps running to clear drilling debris from the bore. Suddenly an apparent blockage gave way, allowing water to flow at its full potential. When they measured the output it was 12 gal. a minute.

GRAY WATER HEAT RECOVERY

Gray water heat recovery is a way of capturing heat that usually goes down the drain. The hardware is fairly simple, consisting of a length of copper drain line wrapped in copper tubing. Savings can be significant.

There are two basic approaches. In the simplest version, drain water heats incoming cold water before it goes to the water heater or to where it's being used. This system works only when hot water is being drawn at the same time it's being discarded, when someone is taking a shower, for example. Although these systems are less expensive, they can't capture heat lost in drain water from a tub, dishwasher, or washing machine because water is not being drawn at the same time.

A more complicated system includes a dedicated storage tank. Drain water flows through a heat exchanger inside the tank and warms up the water. Another heat exchanger at the top of the tank preheats water before it goes to the regular water heater or to where it will be used. Heat-recovery systems without storage tanks are much more common because of their lower cost.

A drain-water heat-recovery system is a simple but effective way of capturing latent heat that would otherwise disappear with the hot water from a shower or washing machine.

Case Study

The Next West House

A superinsulated shell and high-capacity photovoltaic array are all part of the plan

For New Orleans resident Bruce Oreck, Hurricane Katrina brought the challenge of global climate change into sharp focus. When he moved to Boulder, Colorado, he committed to building a home that would contribute no greenhouse gases to the atmosphere and one that could become a zero energy demonstration house for the community. Oreck hoped the design could be replicated anywhere in the Denver metro market, particularly in existing neighborhoods. He hoped it would be the most energy efficient house in the country.

SPECIFICATIONS

House size: 3,617 sq. ft.

Average heating degree days: 5,577

Average cooling degree days: 736

Wall and roof construction: ICFs for basement; SIPs for main floor; 2×6 framing for second floor; 12-in. truss joists for roof

Insulation type/R values: Foundation, R-20 ICFs; walls, R-50 closed-cell polyurethane foam on second floor; R-47 SIPs for first floor; roof: R-60 10-in. urethane foam sprayed

Windows: Loewen double-glazed, low-e, with wood frames; U-0.25, solar heat gain coefficient 0.33 on east, west, and north elevations, 0.50 on south elevation

Photovoltaic capacity: 10kW

Heating source: Geothermal

Air-conditioning: None

Cost per square foot: Not available

Designer: Jim Logan Architecture, Boulder

General contractor: Hughes Construction

The house's solar panels, barely visible from the front of the house, are seamlessly crafted to look as though they are part of the roof. The exterior bricks are recycled from a Denver building.

Oreck hired architect Jim Logan, a long-time advocate for green building and solar design. In part because of Logan's efforts, Boulder now has one of the toughest energy codes in the country, requiring new homes to be 50 percent more energy efficient than the 2006 International Energy Conservation Code (IECC). Logan and Oreck made every decision in the design process–from how to insulate the basement all the way up to the final finishes–after pushing themselves to consider every possibility and the carbon ramifications of each choice.

The result, the Next West House, located in one of Boulder's older neighborhoods, had to meet historical review board standards in addition to a zero energy goal. Solar panels, for example, could not be visible from the street. The finished house met all the goals and then some: Upon completion, the home received a Leadership in Energy and Environmental Design (LEED) Platinum certification and a City of Boulder Historic Preservation Award. The house produces far more power than it consumes, and, over time, it will actually repay the grid for the energy embedded in its construction.

THE PROOF IS IN THE DEMONSTRATION

Although zero energy homes have been built for a number of years, for most people, this idea is largely new or even unheard of. Many of the homes featured in this book have been used as demonstration houses by their developers as a means of educating the local community about this unique way of building. These builders and architects are passionate about their work, and the best way to share this knowledge with the general public is to let them see and experience it for themselves.

When people walk through a zero energy home, it breaks down preconceived notions that green homes are ugly and won't fit in with the surrounding neighborhood. The ability to hear firsthand from a homeowner about their low energy bills is far more convincing than reading about these different building concepts in a journal. These houses demonstrate to the public that a net zero energy home is readily achievable with little additional cost or expertise and will result in huge savings on their energy bill.

Job 1: Superinsulate

Because the basement would be used as living space, construction began belowground with insulated concrete forms (ICFs). Walls are a mix of R-50 structural insulated panels (SIPs) on the first floor and stick framing insulated with INSULsoy closed-cell polyurethane foam (R-6 per inch) on the second floor. Bruce Oreck believes in what he calls "super insulation"—that is, using ICFs, SIPs, and closed-cell insulation combined with tight construction techniques to achieve the almost unheard of score of 0 on the Home Energy Rating System.

The roof insulation for the house is R-60 to R-90, depending on roof orientation; windows are gas filled, low-e squared, with a U value of 0.25. The intention was to insulate the house to the greatest possible extent, to minimize energy loads and allow a variety of supplemental heating sources.

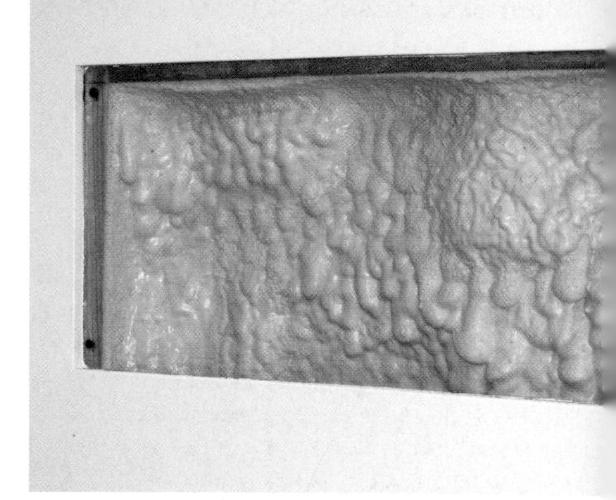

The walls are insulated with closed-cell polyurethane foam. In the garage, a "truth window" cut in the drywall is used to demonstrate this insulation technique to visitors.

And the Winner Is . . . PV

Logan's team analyzed the costs and benefits of a variety of approaches to heating, cooling, and domestic hot water, including active solar, passive solar, thermal solar, and PV. Their spreadsheets included everything from energy loads to electrical usage, solar potential, and building characteristics, and their analysis showed various options based on carbon dioxide emissions. They initially planned for a giant solar hot-water system but eventually settled on a high-capacity PV installation and a ground-source heat pump.

Power comes from 10kW of PV cells that are distributed around the roof to be as unobtrusive as possible. Roof pitches were designed to mount the panels at the best average of summer and winter sun angles for year-round power.

As a net zero carbon house, the Next West House not only will create its own power but also will produce energy to sell back to the grid—enough electricity to pay for all the energy that was expended in building the house and to fuel electric cars in the future.

PV VS. SOLAR: WHAT'S THE DIFFERENCE?

Photovoltaic panels harness the sun's energy to create electricity, which can be used throughout the house in conjunction with electricity from the grid. The size of the array (or panel configuration) depends on the home's typical usage and desired electrical output. A PV system can be expensive, costing anywhere from $10,000 to $80,000, though the cost can be offset by 30 percent from federal rebates and some state tax credits, and excess power can usually be sold back to the grid. It is very likely that the cost of PV will fall in the next few years due to increased demand.

A solar thermal system is used to heat domestic water. Water is pumped through the panels, is warmed by the sun, and is then used for showering or other purposes. Another type of solar system involves pumping a substance like glycol through the panels. The heated glycol is then cycled through the hot-water tank to heat the domestic water. A solar thermal system usually costs less than a PV system, between $3,000 and $10,000.

Green Features Are Key

For Oreck, energy self-sufficiency alone wasn't enough. As he says, "Sustainable building practices really mean the creation of a building that not only maximizes comfort and minimizes energy consumption today but that will also last and adapt to our future living patterns." The Next West House was designed to demonstrate that green building practices are not about sacrifice but about intelligence. Luxury does not have to be in conflict with sustainability.

The proof takes only minutes to grasp. The passive solar–heated, open interior feels larger than it actually is, with high ceilings in the foyer flooded with natural light and an FSC-certified walnut hardwood floor that flows continuously throughout the house. Electrical lighting, which uses chromatically corrected and dimmable LED bulbs that have a color temperature similar to incandescent bulbs, feels merely complementary to natural light. On the second floor, light tubes illuminate the interior bathroom and hallways.

According to Oreck, "Homes need to be built as much for the future as for today. The Next West House incorporates a variety of clever and unique approaches to allow it to flex and change as future technologies and living styles present themselves." Even the walk-in pantry is superinsulated so that if at some point in the future long-term food storage is necessary, the pantry can be retrofitted to serve as

Only FSC-harvested woods and low- and no-VOC paints are used in the house. An induction cooktop can boil water faster than gas or electric and uses less energy.

The faucets provide hot water almost instantaneously, which reduces the amount of water that is typically wasted waiting for hot water.

a walk-in cooler. The water system anticipates the arid future predicted for the Front Range of Colorado. The lowest possible water flow rates for every fixture in the house were selected. The dishwasher and washing machine are the most water efficient available. All toilets are dual flush. Even with four and a half baths, the house will use less than 50 percent of the water of a typical, similarly sized house. With gray water recycling, consumption is halved again, resulting in roughly 25 percent of the water use in a conventional house.

The Mechanical Systems Are Equally Sophisticated

The mechanical room is a building geek's dream. The manifold for the PEX plumbing system is the first thing you see. The "home-run design" means that one fixture at a time can be turned off for repairs, and the plumbing runs to each fixture are kept as short as possible. The hot water at sinks and showers is available almost instantly. The gray water tank looks like a water heater but is self-contained to provide nonpotable water for toilet flushing and meets local plumbing codes.

A gray water system allows the home to use 50 percent less water than the average U.S. home. It captures, filters, and repurifies gray water, which is then automatically used for flushing toilets.

The mechanical room is a far cry from crowded, dingy basements of old, with plenty of room to work on equipment. The stack of batteries at far right is part of the PV system.

The closed-loop ground-source heat pump draws water of between 50°F and 55°F from three wells drilled 200 ft. below the house, allowing it to operate at maximum efficiency for both heating and cooling using the ground water temperature. The home uses radiant-floor heating, which is coupled with a heat recovery ventilation system in the mechanical room that preheats water before it goes into the domestic hot water tank, increasing efficiency.

The entire house is ventilated with a conventional HRV with a variable-speed motor. It is connected to all exhaust fans replacing the evacuated air and runs at low speed on a timer to provide fresh air to the house (because it is so tight).

The geothermal heat pump heats domestic hot water.

There are two main lessons to be learned from this experiment in rethinking American housing. First, super insulation and load reduction strategies make it possible to reach a zero energy goal. Second, if the loads are reduced sufficiently, PV with a ground-

source heat pump is more cost-effective in this climate than is solar hot water. As a result, the family that lives in the Next West House will have a tremendous advantage over their neighbors when energy and water become challenges in Colorado.

Beauty and luxury can go hand in hand with sustainable building practices. In the past, many green homes compromised comfort and beauty for science, but not any more.

Case Study

High-End Lakeside Retrofit

Gutting a 1970s ranch, adding photovoltaic panels, and super-insulating the shell yield net zero performance

A stone's throw above pristine Squam Lake in central New Hampshire, Jane Bindley's 1970s ranch house had a drop-dead view but the kind of energy efficiency you'd expect in a building more than 30 years old. In other words, low energy consumption would never be the point of living there.

SPECIFICATIONS

House size: 5,000 sq. ft., 3,400 conditioned

Average heating degree days: 6,600

Wall and roof construction: Structural insulated panels over original framing

Insulation type: Spray-in polyurethane foam, structural insulated panels

R-values: Foundation, R-42; walls, R-52; roof, R-73

Windows: Triple-glazed

Photovoltaic capacity: 7.5kW

Heating source: Ground-source heat pump

Air-conditioning: None

Cost per square foot: $400

Designer: Ben Southworth

General contractor: Garland Mill Timberframes

Completely overhauled during a 16-month renovation and energy upgrade, this house overlooking Squam Lake in New Hampshire is a net zero energy consumer. Its building envelope and mechanical systems are a marvel of efficiency, but the designers didn't overlook aesthetics. At right, the original kitchen.

Bindley set out to change that. After buying the property late in 2006, she hired Garland Mill Timberframes to undertake a complete overhaul. The Lancaster, N.H.-based company stripped the inside of the house to the frame and rebuilt it. Now Bindley has a Leadership in Energy and Environmental Design (LEED) Platinum weekend and vacation house that uses no fossil fuel whatsoever and meets all of its energy needs with solar energy.

Goal 1: Reduce Energy Loads to Absolute Minimum

For designer Ben Southworth of Garland Mill, and the consultants and subcontractors who worked on the project, the first priority was to reduce energy loads to the barest possible minimum.

After the house had been stripped of interior walls and insulation, it got new wiring and mechanical systems. Then wall and roof

OPEN-CELL VS. CLOSED-CELL FOAM

Polyurethane foam is high-performance insulation that can be sprayed in place, formed into rigid sheets, or used as the core in structural insulated panels (SIPs) between outer layers of oriented strand board (OSB) or plywood.

Foam comes in open- and closed-cell varieties. They are made with different types of gases, called *blowing agents*, which create tiny bubbles that give the hardened foam its insulating qualities. Of the two, closed-cell foam has a higher R-value, about 6 per inch, and also is an effective vapor barrier. It's also denser, roughly 2 lb. per cubic foot.

Open-cell foam, the type used in most SIPs, also can be sprayed in place, has an R-value of about 3.5 per inch, and does not block the movement of water vapor. That can be a consideration when sprayed-in place foam is used in a roof. Open-cell foam has a density of about 0.5 lb. per cubic foot.

Sprayed-in-place foams are especially good at filling hard-to-reach places inside walls and ceilings and forming a good air barrier. In this respect, they are much more effective than fiberglass batt insulation. Foam, which is applied with specialized equipment, is also considerably more expensive.

cavities were filled with spray-in closed-cell urethane foam. On the outside of the house, the company installed 6½-in.-thick structural insulated panels. The result was a wall R-value of 52, nearly three times as high as current recommendations from the U.S. Department of Energy for that climate zone.

The original flat ceiling was removed and replaced with a cathedral ceiling, and closed-cell polyurethane foam was sprayed between the 2×8 rafters. On top of the roof are the same type of SIPs that had been used on outside walls. Together, these two steps brought roof insulation to R-73.

Rafter bays were sprayed full of polyurethane foam then capped on the outside with structural insulated panels, raising the total R-value of the roof to 73.

Raise high the roof beam: The flat ceiling in the original ranch was removed and replaced with a cathedral ceiling supported by a structural ridge beam. The job was a piece of cake for the timberframing company that renovated the house.

A TIGHT BUILDING ENVELOPE

Designers for the Squam Lake project combined spray-in foam insulation and structural insulated panels. The result is extremely low air leakage rates and high insulation values.

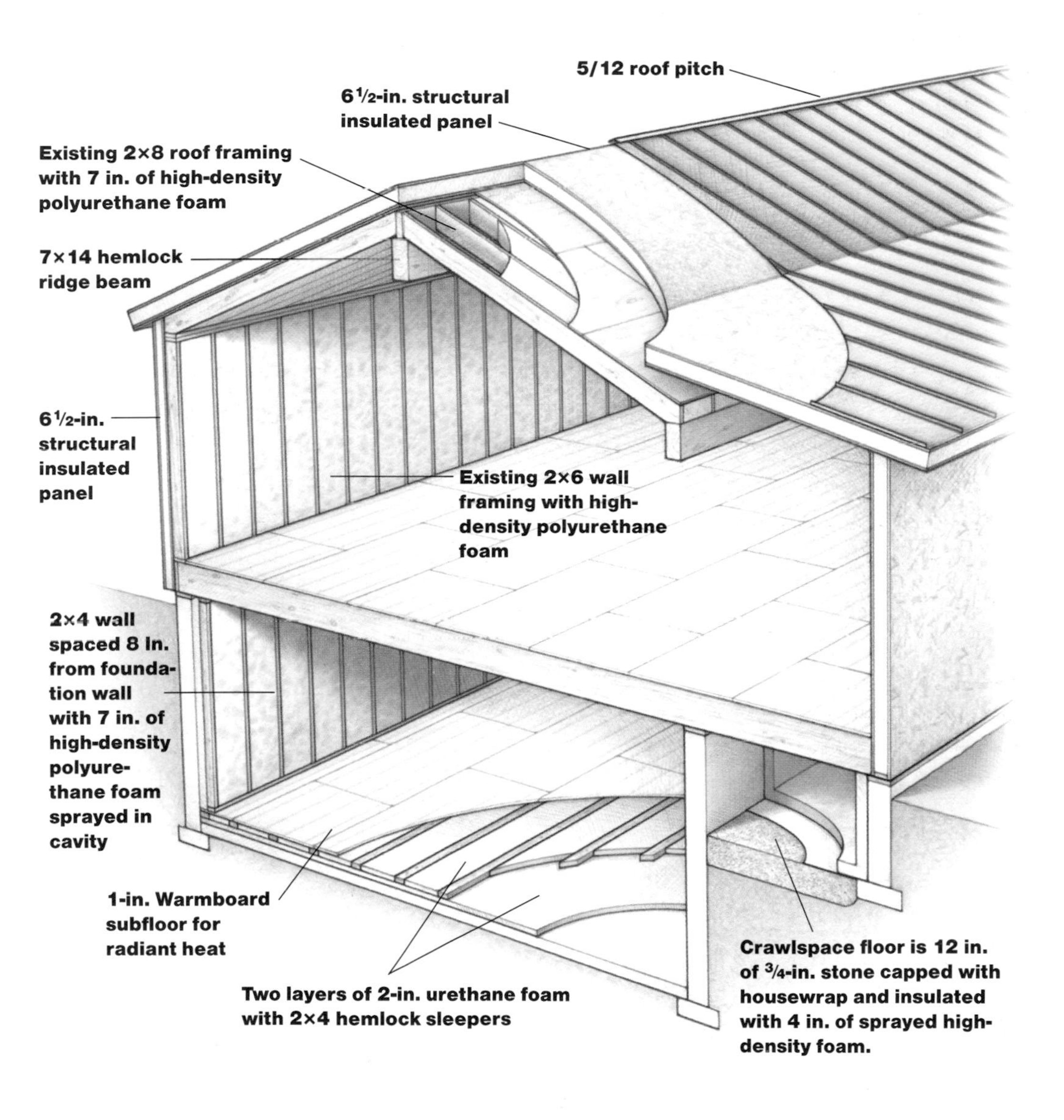

In the basement, Southworth insulated the floor with 4 in. of rigid foam. Then crews framed 2×4 walls and placed them so there was a 4-in. gap between the concrete foundation wall and the back of the framing. With both the framed wall and the cavity behind it filled with polyurethane foam, the R-value climbed from essentially nothing to R-42. Southworth also had foam sprayed between joists at the perimeter of the basement to further reduce heat losses at the rim joist.

Although both the SIPs and the spray-in foam provide an effective air barrier, Garland Mill sealed all seams in the plywood roof and wall sheathing with peel-and-stick tape to provide another layer of protection. Closed-cell foam creates its own vapor barrier so nothing more was added on interior walls.

Windows and glass doors, manufactured by Ottawa-based Thermotech windows, are triple-glazed with two low-e coatings and a low-conductivity gas fill. Frames are fiberglass extrusions filled with insulating foam to help the window reach an R-5 rating (U-value of 0.19). To reduce heat loss through large expanses of glass, the company built moveable window

Taking no chances with air leaks, building designers had all seams between panels of roof sheathing sealed with Vycor® Plus, a peel-and-stick membrane.

insulation from foam-filled panels that could be placed against windows at night and removed during the day.

Garland Mill tested airtightness with a blower-door test and theatrical fog during construction and tightened up any problem spots that were revealed. When the building shell was complete, air leakage had been reduced to 0.03 air changes per hour (ACH) under natural conditions and 0.55 ACH under 50 pascals of pressure. Mechanical ventilation is in the form of a heat-recovery ventilator that takes stale air from the bathrooms and draws fresh air into the bedrooms.

A blower-door test and theatrical fog together helped builders find air leaks in the building envelope and seal them. The result was extremely low air leakage and high energy efficiency.

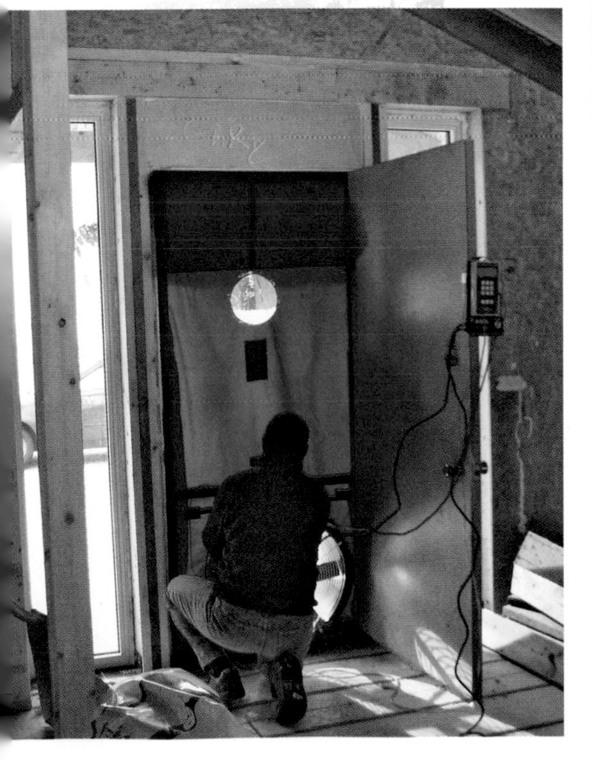

Heat and Power: High Performance at a High Cost

Heat is provided by a WaterFurnace® ground-source heat pump linked to three 200-ft.-deep wells containing closed loops of polyethylene tubing. The heat pump warms water in an 80-gal. insulated storage tank to 95°F, which then feeds a radiant-floor distribution system. Designers incorporated Warmboard, an aluminized structural subflooring in which grooves are machined for polyethylene tubing.

A ground-source heat pump and a radiant floor distribution system keep the house warm through New Hampshire's bone-chilling winters. Flooring was laid over special reflective sheathing in which grooves have been cut for radiant tubing.

The heat pump is part of a heating system that cost as much as $100,000 in materials and labor. If construction budgets had been tighter, Southworth might have favored a different approach—an air-source heat pump with a woodstove backup, for example, or a high-efficiency propane heater. Lower costs for photovoltaic systems also may change the equation in the future.

Domestic hot water is generated by two flat-plate 40-sq.-ft. solar collectors that are connected to two storage tanks totaling 200 gal. The primary tank, equipped with electrical resistance backup heat, is 80 gal.; the secondary tank holds 120 gal. When water on the primary side reaches 140°F, a valve opens and allows water to circulate into the secondary tank, storing the heat for later use rather than dumping the excess.

On-site electricity is generated by a 7.5kW grid-tied photovoltaic system designed by Solar Works, in Boston (now part of Alteris™ Renewables). The $61,000 system includes 36 roof-mounted SunPower® 210w panels, which the manufacturer says are 16.9 percent efficient. The panels generate 15.7w per square foot at peak capacity and are rated for 1-in. hail and 50 lb. per square foot loads.

Using the PVWatts software available through the National Renewable Energy Laboratory, Solar Works had estimated the panels would produce about 9,700kWh a year. But project consultant Marc Rosenbaum told Southworth the number was probably unrealistic because the panels were partially shaded by trees. New calculations predicted 6,800kWh annually.

Doing Is Learning

Bindley invested roughly $150,000 in the heating and photovoltaic systems, about $44 per square foot. In terms of the 16-month

Three hot-water tanks plus all the valves and controls that go with them make for more than your average mechanical room. Solar hot-water panels and a geothermal heat pump are the hot-water sources for the house.

SOLAR PV ROOF ARRAYS

Electricity at the Squam Lake house is provided by 36 rooftop photovoltaic modules. Designers used all of the south-facing roof available to them, and it proved enough to reach net zero energy consumption over the course of a year.

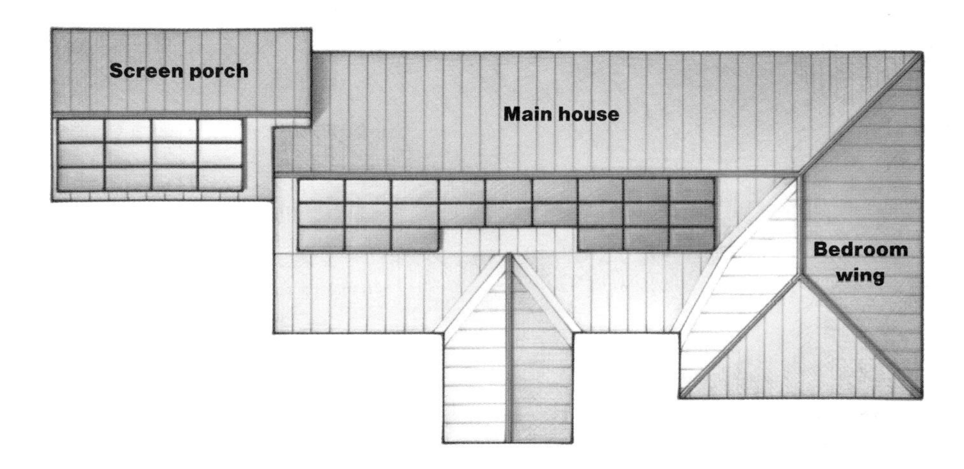

The photovoltaic panels on the roof over the main entry are part of a 7.5kW system.

project total of well over $1 million, the number starts to look less imposing. But the venture is still too expensive to become a template for affordable near net zero housing.

Instead, Bindley has allowed the house to become a kind of open lab, where results can be measured against theory and where specialists continue to tinker with solar and mechanical systems. "I wanted to shoot for a carbon-free house," says Bindley. "That was my intention from the beginning. It's not just an exercise in beautiful home-making." She thinks the money will be well spent if the project is instructive for both builders and homebuyers, and to that end she has opened her doors for public tours and speaks freely about the project.

Aesthetically, Bindley's new house is a huge step up from the dated ranch she bought. From a net zero standpoint, the success of the project turned completely on whether the photovoltaic panels could support the ground-source heat pump and other electrical loads given conditions at the site. On the production side, designers put as many modules on the roof as they had room for, so they kept their fingers crossed, and in the end the project hit net zero energy in April 2009.

For Bindley, it was essential to live in a house that did not contribute to global warming, and she was determined to make the project work. It's not an everyman's house but that's not necessarily the point. "The goal," she says, "is to learn from it."

An early electric bill shows a net balance, not an amount due. Despite snow that blanketed PV panels during one stretch of winter weather, the house showed net zero performance by April 2009.

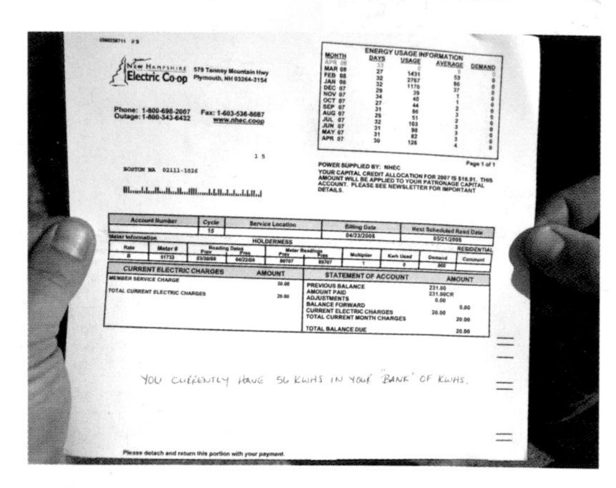

New Hampshire Electric Co-op
579 Tenney Mountain Hwy
Plymouth, NH 03264-3154

Phone: 1-800-698-2007 Fax: 1-603-536-8687
Outage: 1-800-343-6432 www.nhec.coop

BOSTON MA 02111-1026

ENERGY USAGE INFORMATION

MONTH	DAYS	USAGE	AVERAGE	DEMAND
APR 08	33	0	0	0
MAR 08	27	1431	53	0
FEB 08	32	2767	86	0
JAN 08	32	1170	37	0
DEC 07	29	39	1	0
NOV 07	34	40	1	0
OCT 07	27	44	2	0
SEP 07	31	86	3	0
AUG 07	29	51	2	0
JUL 07	32	103	3	0
JUN 07	31	98	3	0
MAY 07	31	82	3	0
APR 07	30	126	4	0

POWER SUPPLIED BY: NHEC Page 1 of 1

YOUR CAPITAL CREDIT ALLOCATION FOR 2007 IS $18.91. THIS AMOUNT WILL BE APPLIED TO YOUR PATRONAGE CAPITAL ACCOUNT. PLEASE SEE NEWSLETTER FOR IMPORTANT DETAILS.

Account Number	Cycle	Service Location	Billing Date	Next Scheduled Read Date
	15	HOLDERNESS	04/23/2008	05/21/2008

Meter Information — RESIDENTIAL

CURRENT ELECTRIC CHARGES	AMOUNT
MEMBER SERVICE CHARGE	20.00
TOTAL CURRENT ELECTRIC CHARGES	20.00

STATEMENT OF ACCOUNT	AMOUNT
PREVIOUS BALANCE	231.00
AMOUNT PAID	231.00CR
ADJUSTMENTS	0.00
BALANCE FORWARD	0.00
CURRENT ELECTRIC CHARGES	20.00
TOTAL CURRENT MONTH CHARGES	20.00
TOTAL BALANCE DUE	20.00

YOU CURRENTLY HAVE 56 KWHS IN YOUR 'BANK' OF KWHS.

Please detach and return this portion with your payment

The object may have been energy self-sufficiency, but the house on Squam Lake also is beautifully detailed and green enough to earn a LEED Platinum rating.

TOP GREEN MARKS FROM LEED

Energy efficiency isn't the only notable feature of Jane Bindley's lakeside house. It's also certified Leadership in Energy and Environmental Design (LEED) Platinum, the highest designation under the LEED for Homes rating system. Some of the sustainable features and materials that Garland Mill used in its construction were:

- **A variety of wood components certified by the Forest Stewardship Council, including the hemlock timberframe, birch kitchen, maple closets, pine trim, and SPF framing lumber.**
- **Salvaged softwood doors and salvaged interior door hardware.**
- **Bathroom tiles made from recycled glass.**
- **Granite slab entry and soapstone hearth from local sources.**
- **Low-VOC paints and sealers.**
- **Clay plaster finishes on walls and ceilings.**
- **A living roof on the two-car garage to reduce water runoff into nearby Squam Lake.**

Jane Bindley's 1970s ranch house was no energy prize when she bought it, but it did have an enviable backyard on the shores of New Hampshire's Squam Lake.

4 Heating, Cooling, and Ventilation

At the heart of net zero performance is a painstaking effort to make the building easy to heat and cool. That means careful air sealing and more insulation than a conventionally built house would have. If energy-efficient builders seem possessed by this process, it is understandable. The U.S. Energy Information Agency estimates that an average of 70 percent of the energy consumed in a U.S. house goes to space heating, air-conditioning, and making hot water. Reducing energy consumption in a net zero house means big payoffs when it's time to select heating, cooling, and ventilation equipment.

By reducing the demand for heating and cooling (what engineers call heating and cooling loads), net zero designs reduce the energy appetite of the mechanical equipment that must be installed. Taken to its practical limits, this strategy makes houses comfortable even without conventional heating and cooling equipment. In the super-efficient PassivHaus designs (see chapter 1), you won't find furnaces at all, just small electric heaters that gently warm the air in ventilation ducts.

The object in net zero building is not necessarily to copy that model but to find a way to keep the house comfortable with the least possible amount of energy. There is no universal solution. Climate is obviously a key factor. In some parts of the country heat is an afterthought; in the middle of a Vermont winter that's about all anyone can think about. Humidity, solar exposure, budget, and how much renewable energy is available on site all play a role in choosing the right equipment.

Passive solar design, which is covered thoroughly in chapter 2, also has a very significant impact on heating and cooling requirements. When solar orientation, windows, and roof overhangs all have been thought through carefully, the need for mechanically generated heating and cooling plummets. Even in very cold regions of the country, passive solar

In this net zero mechanical room, there are no fuel-burning appliances. A compact ground-source heat pump (left) extracts latent heat from a deep well and transfers it to a radiant-floor heating system. It also heats water for domestic use, which is stored in the tank (right). Mechanical systems get their power from a 10kW wind turbine.

CONVENTIONAL HEAT, BIG ENERGY PENALTY

In a net zero building, imported energy must be balanced by renewable energy produced on the site. If the house uses a space heater running on fossil fuels, for example, a wind turbine or photovoltaic panel has to produce the equivalent amount of energy to offset it.

Unless the demand for heat is cut dramatically, that burden can be overwhelming. For example, suppose a house uses 600 gal. of fuel oil per year for heat and hot water. At 139,000 Btu per gallon, that's a total of 83,400,000 Btu, the equivalent of 24,436kWh of electricity.

In a net zero house, a renewable energy source would have to produce at least that much power over the course of a year (let's not forget there also are television sets, lights, and computers to run). In a chilly place like Maine, it would take a photovoltaic array with a peak output of 19kW to make that much energy. If the average price per installed watt is roughly $9 (before rebates and tax credits), a system big enough to handle the job would cost $171,000.

That's why reducing the need for energy is so important.

Passive solar design is an important ally. Floor plans that allow interior spaces to be heated by the sun, along with high mass materials that absorb the energy, mean less reliance on conventional heating equipment and lower energy consumption.

design alone may be enough to keep a house warm and toasty during the day. The house itself becomes a kind of heating appliance.

No heating and cooling option needs to be ruled out categorically in a net zero house. That includes appliances that burn fossil fuels, although mechanical rooms in net zero houses often look dramatically different from those in conventionally built houses. The only caveat is that to reach true net zero performance, the house must produce enough energy from a renewable source (wind, solar, or hydroelectric) to offset the energy content of any fuel that is burned.

If there is one constant, it's that zero energy houses must have some kind of mechanical ventilation. Because net zero buildings are so tight, throwing open a window here and there isn't enough to guarantee good air quality. Ventilation should be engineered into the building at the earliest stages of design to make sure the building and the people who live there stay healthy.

Mechanical systems in a net zero house can be more complex, especially when coupled with renewable energy. Where a conventional house might have a single water heater and storage tank, this one has three, two of them for the solar hot-water collectors and another powered by a ground-source heat pump used for radiant-floor heat.

MAKING THE ENERGY NUMBERS WORK

Designing a net zero house is far from a back-of-the-envelope operation. But it's interesting to play with some numbers for an idea of how much renewable energy would be required to offset fossil fuels.

It's helpful to convert all energy units to kilowatt hours because that's the currency of photovoltaic systems and wind turbines. But we don't buy fossil fuels by the kilowatt. Fuel oil is sold by the gallon, natural gas by the cubic foot. To run some estimates, assume that a gallon of no. 2 heating oil contains 139,000 Btu of energy; natural gas roughly 1,030 Btu per cubic foot (these and other energy conversions are readily available on the Internet). To convert Btu to kilowatt hours, divide by 3,413.

On the production side, a software program called PVWatts developed by the National Renewable Energy Laboratory and available for free on the web will cough up an estimate of how much electricity a PV array of a given size will make in a year. That's a good place to start. Estimating wind production is a little trickier, but the simple formula explained in chapter 3 (see p. 124) is a good starting point with a turbine of a known peak output.

HOW TIGHT IS TIGHT ENOUGH?

Net zero houses are substantially tighter than conventional, code-compliant houses. That much is a given, and it means whole-house ventilation is a necessity.

How tight should energy-efficient houses be? There's no standard definition, but Paul Eldrenkamp, the owner of a custom remodeling company in Newton, Mass., has a starting point. Writing in *The Journal of Light Construction,* Eldrenkamp suggests there should be less than 1 air change per hour in a blower-door test generating 50 pascals of pressure.

Options for Heat

Designers and energy consultants are likely to use computer software to determine how much energy will be needed to keep a house warm in winter. Variables such as insulation, windows, solar orientation, thermal mass, and square footage all have an affect on the outcome. Calculations have traditionally been made with the help of *Manual J,* a guide produced by the Air Conditioning Contractors of America®. But alternatives, such as Energy-10 from the National Renewable Energy Laboratory (see the Appendix 1 on p. 242), may also be an option. Whatever the specific tool, the days of a heating contractor taking an educated guess based on square footage alone are long gone. A seat-of-the-pants estimate just won't work.

What the heat designer needs is a load calculation. This is the energy (expressed in British thermal units, or Btu) the building will lose per hour and, consequently, the amount of energy that a heating system must produce to keep the house comfortable. With this number in hand, a heating contractor or energy consultant can start thinking about specific kinds of equipment. Options include air- and ground-source heat pumps, electrical-resistance elements, appliances that burn fossil fuels (natural gas, liquefied petroleum, fuel oil, or kerosene), and wood or pellet heaters (usually lumped under the heading of "biomass" heaters).

An all-electric house certainly makes the bookkeeping easy. If heating and cooling equipment run on electricity alone, all the homeowner has to do is watch the electric meter to keep tabs on the net zero status. There's no need to convert the energy potential of another fuel (heating oil, for example, or natural gas) into an equivalent kilowatt hours of renewable energy produced at the house to make sure they balance.

In addition to the specific fuel or appliance, designers also have to consider how heat will be distributed around the house: via air ducts, tubing embedded in the floor, baseboard or free-standing radiators, or by natural convection in a house with an open floor plan. There are pros and cons to all of them.

Heat Pumps Provide Both Heating and Cooling

Homeowners and builders alike seem infatuated by heat pumps these days, and probably with some justification. There are two kinds: air source and ground source, both of which can be used for space heating and cooling as well as heating water.

All heat pumps share a technology that may look like black magic to a nonengineer: How is it possible, one might reasonably ask, to heat a house to a comfortable 68°F with air or water that is only 40°F or maybe less?

A heat pump works on the principle of vapor compression; it is simply a refrigerator running in reverse. At the start of the heating cycle, water or air warms up a refrigerant in a heat exchanger, turning the refrigerant from a liquid into a gas. Next, the gas is compressed by an electrically driven pump, which boosts its temperature. When the gas passes through a condenser and then an expansion valve, it gives off heat and eventually returns to a liquid state to start the process over again. The captured heat can be used to warm air or water.

Heat pumps don't burn a fuel directly to produce heat like a furnace; instead, they

VAPOR COMPRESSION DRIVES HEAT PUMPS

Like a refrigerator or air-conditioner, a heat pump relies on the vapor compression cycle to extract heat from air, earth, or water. A closed loop of refrigerant absorbs heat, which is then concentrated by a compressor. The high-temperature gas passes through a reversing valve and then to a condenser coil in a secondary heat exchanger where, in this example, it warms air for distribution around the house. Heat also can be transferred to water and distributed through a network of tubing in the floor. By reversing the flow, a heat pump also can cool the house. A desuperheater can be added to heat water for domestic use.

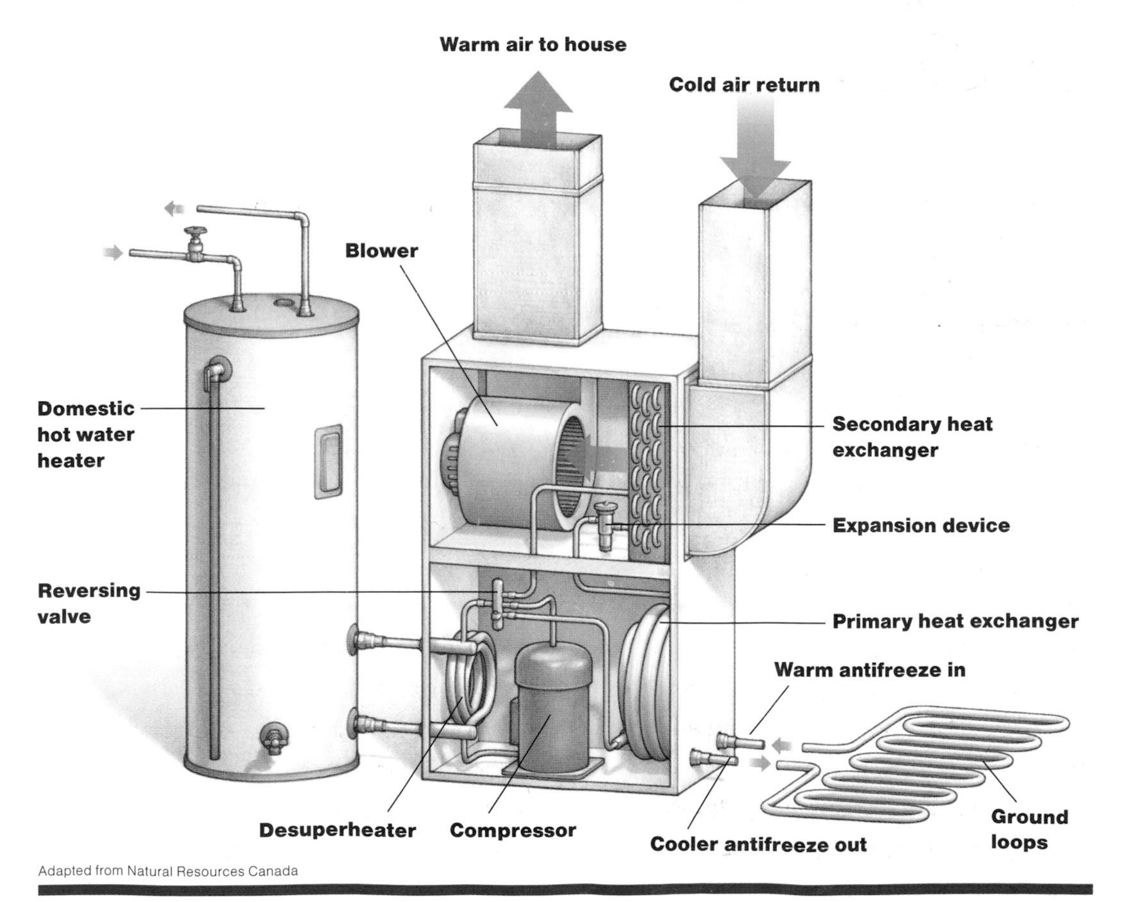

Adapted from Natural Resources Canada

move heat from one place to another using only electrically driven pumps and valves and a closed loop of refrigerant. In summer, they can draw heat from inside the house and dump it, either into the air or into the ground via the same underground loop used in winter for heating. Both air- and ground-source heat pumps can be very energy efficient, although they are rated in different ways.

Ground-source heat pumps are efficient but costly

Ground-source heat pumps, often called geothermal heat pumps, tap into the earth or a source of water for residual heat. In a common scenario, loops of plastic tubing filled with a nonfreezing fluid are buried in trenches that may stretch hundreds of feet around a building site. Alternately, tubing can be dropped into one or more vertical wells, a technique that takes a lot less room but can be quite costly. Tubing can be placed at the bottom of a nearby pond, or water from a well can be the source of latent heat. In other words, there are all kinds of ways to connect the heat pump with the ground.

Extensive trenching and boring deep wells are both expensive, which can make the cost of a ground-source heat pump higher than many if not all other conventional heating options. But two houses detailed in this book provide good examples of how costs can be reduced with some creative thinking.

(continued on p. 173)

COMPARING HEAT PUMP PERFORMANCE

The highest performing ground-source heat pumps on the market offer excellent energy savings over their lifetime—at least on paper. In one simulation from the U.S. Department of Energy, the most efficient heat pump was expected to use only about one third of the energy for heating as a standard air-source heat pump, 10,900kWh vs. 29,800kWh. Savings for cooling were not quite as dramatic, but overall the simulation estimated that annual energy savings were better than 2 to 1 with the ground-source heat pump.

Even so, anecdotal evidence suggests that ground-source heat pumps can be finicky and disappointing. Efficiency is described as a coefficient of performance (COP) which is the heating capacity of the device in Btu divided by the electrical input. A heat pump with a COP of 4, for instance, produces 4 units of heat energy for every 1 unit of electrical energy it consumes. But COPs published by manufacturers aren't the whole story. One reason is that the COP refers to the energy consumed by the heat pump itself, not the electric pumps that move the transfer fluid through all that tubing in the ground.

Ground-source heat pumps also lower the temperature of the soil around the tubing during the winter. The colder the soil, the less efficient the heat pump, and the nonfreezing glycol that must be used where freezing is a threat takes more energy to pump while offering a lower heat-transfer capacity. Some of these factors are extremely difficult to predict.

Ground-source heat pumps are more efficient than conventional electric-resistance heat and air-source heat pumps. But they are not magic. Efficiency can be lower (and thus the consumption of electricity higher) than their COPs would suggest, derailing the efforts of even some very bright energy designers.

Ground-source heat pumps typically rely on hundreds of feet of tubing placed in horizontal trenches to pick up latent heat from the earth. Where this kind of excavation would be too disruptive, tubing can go in one or more vertical wells.

The seeming magic of geothermal heat comes in a pretty small package. This unit, connected to heat transfer tubing in vertical wells, is large enough to heat 3,400 sq. ft. of space in a home in central New Hampshire, where winter temperatures can dip well below zero. It burns no fuel directly.

GROUND-SOURCE HEAT PUMPS

Horizontal Ground-Linked Heat Pump

Tubing filled with antifreeze absorbs heat from the ground and carries it to a ground-source heat pump. Even in winter, underground temperatures are high enough to allow the system to work, but hundreds of feet of tubing must be buried and excavation can be extensive.

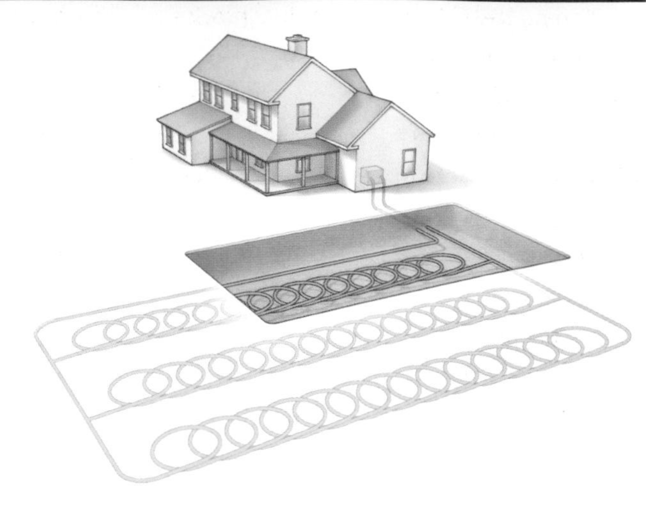

Open-Loop System Where Water Is Abundant

With an open-loop system, water is pumped from a well through the heat pump and returned to the earth. It passes through the system only once. If the well produces enough water, it can be used for both domestic supplies and the heat pump.

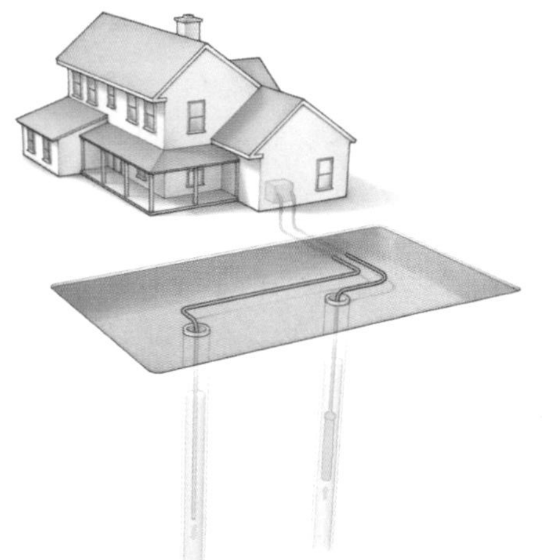

Closed-Loop Vertical System Recirculates Same Fluid

Tubing can be placed in one or more vertical wells if there isn't enough room for horizontal trenches. The number and depth of the wells depends on a variety of factors. A closed loop of nonfreezing liquid absorbs heat from the earth and moves it to the heat pump inside, where it is concentrated and distributed around the house.

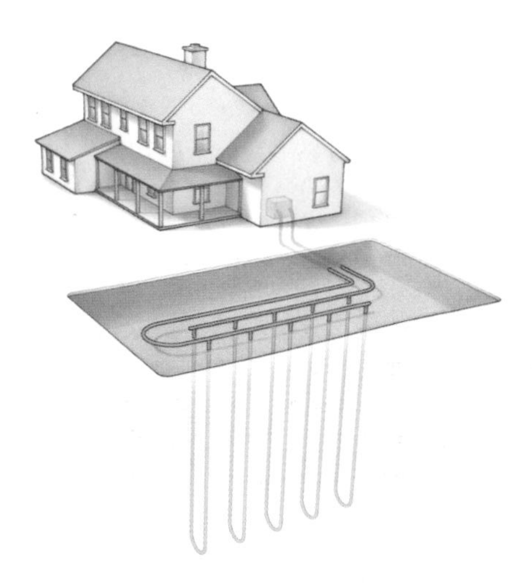

(continued from p. 170)

In Vermont (see "The Good Life in Vermont" on p. 138), the same well that provides drinking water for the family also supplies water for the heat pump. The well provides enough water for both purposes (12 gal. per minute), not a luxury everyone will have but a possibility on some sites. The risk is that over the course of a winter so much heat could be pulled from well water that it would approach the freezing point. To guard against this possibility, the system includes a bypass valve that automatically diverts the water rather than returning it to the well when the well water temperature reaches 34°F.

In a near-zero project in Lenoir City, Tennessee (see "All-Electric House: $1.16 a Day" on p. 210), designers from the Oak Ridge National Laboratory found a way to use trenches that had already been dug around the house for the 1,500 ft. of tubing needed for the heat pump. Because there wasn't any extra excavation, costs were lower and disruption to the lot much less. It took coordination during construction to make this possible, along with some careful calculations to minimize the risk that earth around the house would freeze and damage the foundation.

Ground-source heat pumps are not necessarily prone to more mechanical problems than other systems, but it sometimes seems that way. They are complicated devices and a malfunction that throws off net zero performance can be hard to track down.

Despite the drawbacks of potentially high cost and system complexity, the energy efficiency of a ground-source heat pump is attractive. The Department of Energy estimates that geothermal systems can be as much as 60 percent more efficient than conventional heating systems. Efficiency is measured by something called the coefficient of performance (COP). It's what you get when you divide the Btu of heat produced by the heat pump by the amount of electricity, in Btu, that's used to produce it. So, for example, a heat pump given a COP of 4 by the manufacturer should produce 4 Btu of heat for every 1 Btu of electrical energy it uses.

Not everyone, however, believes the advertised efficiencies (see "Comparing Heat Pump Performance" on p. 170). Henry Gifford, a New York City heating specialist, contends that manufacturers' COP values are overstated because they may not include all the electricity required to power groundwater pumps and circulation fans in warm-air systems.

OPEN-LOOP SYSTEMS ARE MORE EFFICIENT

Ground-source heat pumps with closed loops circulate a nonfreezing liquid through tubing placed in horizontal trenches or in vertical wells. In contrast, open-loop systems make use of water from a stream, well, or lake that passes through the system only once before it is returned to its source or jettisoned elsewhere.

Of the two, the best open-loop systems available are slightly more efficient. According to the U.S. Department of Energy, open-loop heat pumps can operate with a COP of 5.5, vs. 4.9 for the best closed-loop system.

Gifford also points out that utility power is roughly 30 percent efficient—that is, for every unit of energy potential at the coal, oil, or gas generating station about one third gets to the house in the form of electricity. So in the sometimes-convoluted world of net zero, an oil furnace operating at 85 percent efficiency may use less fuel to heat a house than a ground-source heat pump with a COP of less than 3 when it operates on utility power. This is a good place for a competent energy consultant to wade in and look carefully at the particulars of a house and its heating options before deciding what route to take.

An air-source heat pump is simpler, less expensive, and common in parts of the country where air-conditioning is as much of a priority as heat. Conventional designs, however, lose much of their efficiency when air temperatures drop below 40°F.

Air-source heat pumps well suited to mild climates

Air-source heat pumps work in much the same way, except that they use outside air as the source for heat rather than a link to the earth. They're very common in the southeastern United States, where milder climates can make air-conditioning as important as heating, but they are not as attractive in cold climates.

Air-source heat pumps are less expensive than ground-source heat pumps, and they're easier to install because they don't require any trenching or drilling. Most designs, however, begin to lose efficiency as temperatures fall below 40°F. When that happens, electric-resistance coils or a gas-burning furnace kicks in, lowering overall efficiency.

Manufacturers have been working to develop special cold-climate designs that work efficiently in the northern tier of the country. Though they are still relatively new, these spe-

HEAT PUMP REFRIGERANTS ARE GETTING BETTER

Heat pumps use chemical refrigerants in a vapor compression cycle for heating and cooling, and the process is getting more environmentally friendly.

Chlorofluorocarbons are a class of refrigerants (like Freon 113 made by DuPont) that worked very well but were also very destructive to the Earth's ozone layer. They have since been succeeded by hydrochlorofluorocarbons (blessedly abbreviated as HCFCs), a class of refrigerants that do less damage. One widely used HCFC is R-22. This, too, will be phased out in favor of refrigerants that do not deplete the ozone layer, such as a chemical called R-410A. Some manufacturers have already made the switch.

HEATING SEASONAL PERFORMANCE FACTOR

The efficiency of air-source heat pumps is measured with the heating seasonal performance factor (HSPF), which represents the heating efficiency over an entire heating season. The rating is derived by dividing the Btu of heat provided over the heating season by the watts of electricity used over the same period. This number can be converted to a percentage of efficiency by dividing the HSPF by 3.414, the number of Btu in 1W/hr of electricity. A heat pump with a HSPF of 7, for example, has an efficiency of 205 percent. For every Btu of electricity used to operate the heat pump, 2.05 Btu of heat are returned. This would make the heat pump twice as efficient as electric baseboard heat, which is roughly 100 percent efficient.

The HSPF covers the efficiency of the compressor and the electric resistance heater. The most efficient models have an HSPF of between 8 and 10, according to the Department of Energy.

cialized heat pumps should cost significantly less than ground-source models and still offer good efficiency even in below-zero weather. Hallowell International, a Maine-based manufacturer, says its cold-weather design is 200 percent efficient at 15°F below zero. Its Acadia™ heat pump incorporates a second compressor that kicks in when outdoor air temperatures fall below 30°F, making the refrigerant even colder and able to absorb heat in colder air. Although supplemental heat may still be needed, it's activated at a lower point than are conventional air-source heat pumps.

New designs may extend the range of air-source heat pumps from their conventional stronghold in mild climates. This one, manufactured by Maine-based Hallowell International, uses a second compressor to improve efficiencies at low temperatures.

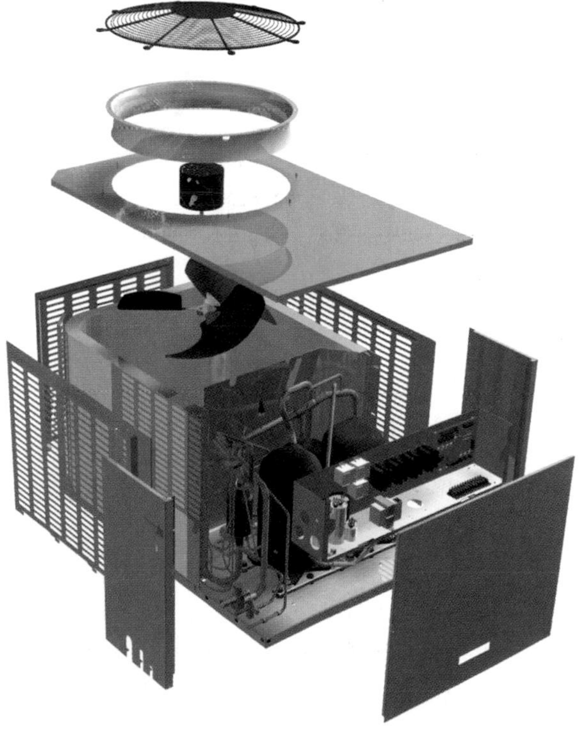

One of the big advantages of an air-source heat pump is that it provides air-conditioning as well as heat. Cooling efficiency is measured by the seasonal energy efficiency ratio (SEER); the higher the number, the more efficient the device. Heat pumps with SEER ratings of 18 are available.

Installation is critical to performance. Not only should ducts be tightly sealed with mastic to prevent leaks and pressure imbalances inside the house but also the system needs to be charged with the right amount of refrigerant if it's going to work as designed. The Department of Energy, citing an EnergyStar study, suggests that 50 percent of all air-source heat pumps are installed with leaky ducts, the wrong refrigerant charge, or too little airflow. Gifford suspects the numbers are even higher. Heat pumps charged in the field are more susceptible to problems than those charged with refrigerant at the factory.

Electric-Resistance Heat Can Work in a Net Zero House

Electric heat has the reputation of being an expensive heating option, and because it has no cooling mode it may be a second choice in climates where mechanical cooling is required during the warmest parts of summer. But electric heat has some advantages, too, and it can be used successfully in a net zero house.

The extremely low heating loads in a superinsulated house change the rules. In the net zero Habitat for Humanity house in Wheat Ridge, Colorado, for example (see "Affordable Housing Meets Zero Energy" on p. 233), heat is provided by electric baseboard heat in the bedrooms and a gas-fired space heater in the living room. With a moderately sized photovoltaic system (capacity of 4kW), the house was able to show a net surplus of energy over a 1-year period when measured on a source basis.

REVERSE-CYCLE CHILLERS ARE MORE FLEXIBLE

One of the problems with a conventional air-source heat pump is that it will occasionally have to run in reverse during the heating season to keep the evaporator from freezing. When that happens, cool air is distributed around the house. A type of heat pump called a reverse-cycle chiller eliminates this problem by heating a tank of water as it runs and using the water to defrost the evaporator instead of warm air from inside the building.

Water is stored in a superinsulated tank. During the heating season, it can be used with a fan coil to circulate warm air through the house, but the water also can be used in tandem with a radiant floor distribution system. In summer, the heat pump cools the tank of water and uses it to cool the house.

Reverse-cycle chillers also can be used to heat domestic hot water in both summer and winter. These features make reverse-cycle chillers somewhat more expensive than a standard heat pump.

Electric heat is 100 percent efficient. All, or virtually all, of the electricity the system uses goes directly into heat, compared with 85 percent to 95 percent of the energy potential for high-efficiency gas or oil furnaces and boilers. It's also much less expensive to install than a heat pump, and it can easily be zoned on a room-by-room basis so rooms that aren't being used don't have to be heated at all. Electric heat saves the cost of a chimney, and it produces no combustion gases so it's safer than any fuel-burning appliance.

Although cost is not necessarily part of the net zero equation, electric heat can be competitive with a conventional boiler or furnace that burns a fossil fuel. At the national average of about 11 cents a kilowatt hour, electric heat is cheaper when fuel oil costs $3.80 or more per gallon. Oil was substantially more than that in mid-2008 when it topped $4.70 in parts of New England. It is likely to cost that much again when the world economic slowdown eases and the demand for oil goes back up.

Electric resistance baseboard heaters are inexpensive, easy to install, and make it simple to zone heat. When heating loads are very small, the heaters are an option in a net zero house.

Radiant wall or ceiling panels

Electric baseboard heaters are inexpensive and easy to install, and as a result they're very common in houses where electric heat is used. But they're by no means the only choice. Electric radiant panels mounted on either walls or ceilings are another option.

Radiant panels heat objects directly, not by heating an intermediate medium in the room (air) that in turn transfers heat to people. People can feel comfortable at lower temperature settings than would be required with a conventional heating system.

Surface-mounted radiant panels are low in mass, meaning they can heat up quickly, and as thin as 1 in. so they aren't obtrusive. Ceiling-mounted panels have one advantage in that they distribute heat without interference from furniture or other objects in the room. Some panels incorporate resistor wires in 1/2 -in.-thick gypsum drywall. The panels are installed between ceiling joists and then capped with finish drywall, so the radiant panel is completely invisible.

A study in 1994 by the National Association of Home Builders Research Center found

HIDDEN HEAT IN THE CEILING

Electric elements installed in the ceiling and capped with drywall provide radiant heat but are completely out of sight. Surface-mounted wall and ceiling panels also are available from several manufacturers.

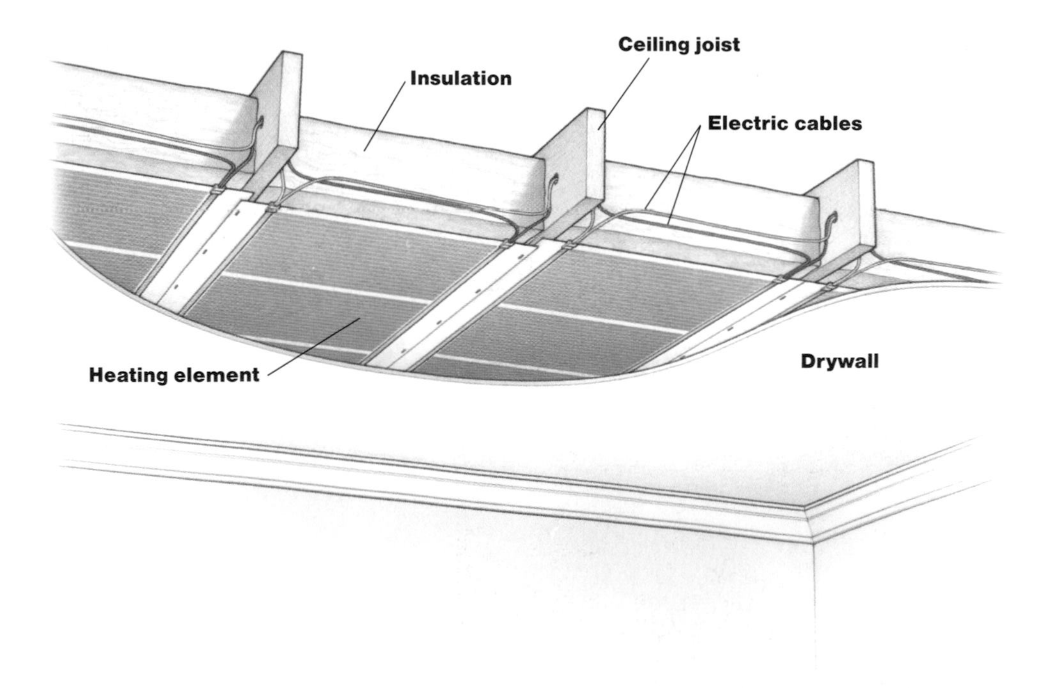

that radiant panels were more efficient than either electric baseboard units or an air-source heat pump. The study was very limited in scope and so it may be dangerous to draw any sweeping conclusions about the relative merits of radiant panels to other types of electric-resistance heating. But the numbers are attractive enough to warrant a comparison when electricity is the heating fuel of choice.

Electric grids that go in the floor

Electric heat also can be in the form of a mat or grid of wires that's buried in a layer of thinset mortar and covered with tile, stone, or engineered wood flooring. Like other varieties of electric heat, floor grids are easy to zone and they provide even, radiant heat.

Electric floor grids are most often used in a bathroom or kitchen or another spot where a boost of extra heat is needed periodically. But when connected to a programmable thermostat, these radiant systems also can take over for a conventional central heating system. The advantage is that heat is provided only when and where it's needed. Another advantage is that tile or stone floors can be warmed with electricity purchased at off-peak rates, lowering overall heating costs (but not, sadly, overall consumption).

Overall, electric-resistance heat does have some pluses, chiefly the ease of zoning heat and lower installation costs compared with conventional central systems. What's working against it from a net zero standpoint

Mats containing a grid of electric resistance cable can supply radiant heat. The mats are embedded in a layer of thinset adhesive and covered in stone, tile, or even engineered wood flooring and can be controlled with a programmable thermostat to take advantage of off-peak electric rates. These are made by Nuheat™.

is that it's not as efficient as a heat pump, particularly a ground-source heat pump. A heat pump squeezes more benefit from each kilowatt hour of electricity, and keeping energy use to a bare minimum is the overriding goal in net zero design. Even so, electric-resistance heat can be part of a zero energy package. In parts of the country where electric rates are low, it can be attractive from an operating cost standpoint, too.

TAKE ADVANTAGE OF OFF-PEAK RATES

Floors with high thermal mass that are heated with embedded electric cables can be slow to warm up. Slow response is a disadvantage in a room that's going to be used for only a few minutes. But a floor that is slow to warm also is slow to cool off. Taking advantage of lower off-peak electricity rates with a programmable timer is one way to save money (although not necessarily kilowatt hours).

Wood and Other Forms of Biomass

Wood heat is a familiar standby in rural areas where sources of high-quality cordwood are plentiful and prices relatively low when compared with other fuels. Many of the wood stoves, boilers, and furnaces manufactured today are much more efficient and less polluting than those made a generation ago. Tougher federal air-quality regulations are one reason, and the technology of clean-burning wood appliances has come a long way.

Where firewood is plentiful, a wood stove can be a cheerful companion in winter. Stoves are far more efficient (and more stylish) than they used to be, but they still emit combustion particulates and need regular maintenance.

Pellets are compressed chunks of wood fiber that burn with less residue than wood. They can be ordered by the ton and stockpiled more easily than cordwood.

Pellet stoves have more automatic features than wood stoves, including hoppers that automatically feed the combustion chamber. But they won't run when the power goes unless they have a battery backup. Ecotek™, which manufactures this stove, claims an efficiency of 84 percent.

Wood may be cheaper than other fuels and it can be sustainably harvested, but it's not considered a renewable form of energy like solar or wind by some zero energy advocates.

On the plus side, wood can be an infinitely renewable resource that's locally harvested. A landowner who manages a wood lot carefully can harvest a certain amount of wood indefinitely without degrading the health of the forest. An old rule of thumb, at least in New England, is one cord (a pile measuring 4 ft. by 4 ft. by 8 ft., or 128 cu. ft.) per acre per year.

There are wood furnaces, wood boilers, and freestanding wood stoves. Any of them is capable of heating a whole house. Appliances also can burn corn or compressed wood pellets, and some multifuel appliances can burn one of two fuels (wood or fuel oil, for example) depending on which is more available or cheaper.

Cutting, hauling, and stacking wood is good exercise, and it embodies all the wholesome self-reliance of an L.L.Bean® catalog. The problem is that burning wood inevitably releases particulates and other contaminants into the air, so from an emissions point of view it's not as desirable as photovoltaic panels or a wind turbine, which makes electricity without an environmental penalty. In addition, biomass heaters are not automatic like heat pumps, electric heaters, or fossil-fuel heaters. Someone has to load the fuel, keep the firebox clean, and tend to the device on a very regular basis. Chimneys must be inspected and cleaned annually, and there is always a risk of fire. On a net zero basis, there also is some tricky energy accounting that goes along with biomass, not all of which is necessarily going to be accurate.

This is especially true with wood. There are no standards for the energy content of firewood. A cord of hardwood could contain the energy equivalent of between 15 million and nearly 30 million Btu. That's a very big swing. Wet wood doesn't burn nearly as well as dry wood, and unless the buyer uses a moisture meter to test each piece of wood on the way to the stove there is no way of knowing its water content. Wood may be sold as "seasoned" or "dry" but never as "no more than 20 percent moisture content."

With that in mind, how much electricity must be produced on site to compensate for the energy contained in the firewood, thereby reaching net zero performance? If the homeowners got their hands on a dry cord of white oak they would have the potential of about 25 million Btu, the equivalent of about 7,300kWh of electricity. But some deduction must be made for the stove because it is not

100 percent efficient, and there's no telling if the estimate of embodied energy in the wood is accurate to begin with.

David Barclay, then executive director of the Northeast Sustainable Energy Association, called the use of wood and other forms of biomass for heat "a fuzzy issue." The organization prefers cleaner forms of energy. Wood and biomass, he said, produce carbon as they burn so if the goal is clean energy that's going to be a problem. "In the end," Barclay said, "I think you have to go with the standard definition of [zero net energy]—that the building has to produce more energy than it uses. Suppose someone could get all the power/heat from burning biomass. Would they be ZNE? We would say no."

That may well be true. And it wouldn't take too many cords of wood to create a staggering kilowatt hour deficit that a PV array or wind turbine would then have to make up. That said, we think sustainable, small-scale wood harvesting is infinitely preferable to ripping off Appalachian mountaintops and extracting coal to burn in inefficient power plants.

Consider an all-electric house with a ground-source heat pump, an electric water heater, and a big photovoltaic array to power everything. In one sense, it's a very clean operation. Nothing is burned on site, so the house produces no greenhouse gases directly. But if the house is tied to the utility grid, the electricity that it imports could very well be made in a coal-burning plant. Moreover, the power is only 30 percent efficient so we have to burn roughly three units of energy to get one unit at the house. Putting zero energy considerations aside for the moment, which of these options looks better?

Getting Heat Around the House

Producing heat is only half the battle. It obviously must be circulated through the house before it's of use to anyone. Passive solar designs that encourage the movement of air by convection are very appealing, but most houses will need a mechanical system of some kind and that amounts to choosing between air and water as a transfer medium.

Forced-air systems are more common, mainly because they cost less to install than hydronic heat. Another advantage is that the same ducts that carry warm air in winter can be used to cool and dehumidify the house in

High-efficiency furnaces squeeze as much as 95 percent of the potential energy from natural gas. One problem with conventional heating appliances, however, is that they are often oversize for a house with extremely low heating loads.

summer. High-efficiency filters can clean pet dander, dust, and other irritants out of the air, making the house more comfortable and better for health. Hydronic systems can't do that.

The problem with forced-air distribution systems is that ducts may not be sealed carefully enough. Leaky ducts can cause a variety of problems, including pressure imbalances that encourage air leaks through the walls and, as a result, lower heating and cooling efficiency. Ducts that run through overheated attics or chilly crawl spaces also waste energy, and losses can be significant. Sealing ducts with mastic (not duct tape), and keeping them inside the conditioned space, is essential for anyone hoping to get to net zero performance.

Radiant-floor distribution uses cooler water

In a conventional hot-water system, a boiler heats water to 160°F or more before it is pumped through baseboard or freestanding radiators. Radiant-floor systems are fundamentally different. Water is distributed through a network of plastic tubing in the floor, warming people and objects in the room directly by radiation. A room can be comfortable with a lower thermostat setting than would be possible with a more conventional heating system.

More to the point, the water doesn't have to be as hot as it does in conventional hydronic radiators. Even in the dead of winter, outgoing water might not have to be more than 100°F to keep the house comfortable. When the heat source is a fuel-burning appliance, a sensor called an outdoor reset can be used to regulate water temperature based on outside conditions. When demand is low, the water temperature is correspondingly lower. Lower water temperatures also mean that solar panels can be used to provide, or at least supplement, the hot-water supply. When radiator water has to be 160°F or higher, solar panels can't keep up.

Radiant-floor tubing can be installed under a variety of materials, including engineered wood. High-mass floors, such as concrete,

Ducts should be sealed with mastic, not duct tape. Leaky ducts are a common problem that waste significant amounts of energy and can lead to a variety of issues, especially when they run through unconditioned spaces.

A hot-water radiator requires higher water temperatures than radiant-floor heating systems, making radiators a poor fit with solar hot-water collectors.

CHOOSE FLOORING CAREFULLY

Radiant-floor heating is an ideal partner with tile, stone, or concrete because the high thermal mass of these floor coverings moderates sharp swings in temperature. Wood flooring also is a possibility, but some caution is necessary.

Wide plank floors that are flat sawn—that is, with growth rings tangential to the face of the board—are more likely to shrink, expand, and cup when exposed to cyclical changes in temperature. A better choice is engineered wood, such as bamboo, or quarter-sawn wood where the growth rings are perpendicular to the face of the plank. They're much less susceptible to dimensional change. Narrow boards are less likely to be affected than wide ones.

tile, or stone, help moderate temperature fluctuations: They are slower to heat up, but retain heat longer. And the distribution system is invisible. There are no radiators or baseboard units to arrange furniture around.

A number of houses profiled in this book use radiant-floor distribution systems, usually in conjunction with a ground-source heat pump. One drawback is cost. On top of the costs associated with installing the heat pump itself, running all that tubing in the floor is labor intensive and expensive unless the owner/builder installs it himself or herself. Although it is possible to adopt this heating strategy on a modest budget (as in the Habitat for Humanity house in Lenoir City, Tennessee), these systems more often seem to be associated with high-end projects for which money is not as tight.

Radiant-floor tubing placed in a floor with a lot of mass, like concrete, provides gentle heat and helps moderate swings in indoor temperatures. Water can be much lower in temperature than what's needed for conventional hot-water radiators. Unlike air registers, radiators, or other forms of heat, radiant-floor tubing is completely invisible once installed.

Radiant-floor tubing also can be installed in a low-mass floor. In this case, PEX tubing is fitted in specially designed subflooring with an aluminum backing that helps reflect heat upward. Engineered wood flooring makes a good finish surface.

Keeping the House Cool

In many parts of the country, sultry summer weather lasts only a couple of weeks, not long enough to warrant a complicated cooling system. That's not the case in parts of the desert Southwest or the humid South where suffocating heat, high humidity, or both make parts of the year unbearable without air-conditioning.

Here, too, sensible passive design can make a world of difference. At an early near-zero house in Lakeland, Florida (see "High Heat" on p. 42), researchers from the Florida Solar Energy Center found that extended roof overhangs, reflective roof coverings, smart duct design, and keeping the mass of masonry walls inside the thermal barrier all contributed to substantially lower energy demands during the cooling season. At the very least, the design offers a strategy for lowering peak electrical demands for the local utility and avoiding the construction of new power plants.

Opening windows at night, turning on a whole-house fan to flush out hot air and bring in cool nighttime air, and then closing windows first thing in the morning can be extremely effective, especially when the house is superinsulated. In areas where central air-

GETTING TO ZERO

An attic fan that flushes stale air out of the house at night can be a very economical way to cool down the house. Closing windows in the morning keeps cool night air inside. In a well-insulated house, a whole-house fan may make conventional air-conditioning unnecessary, at least in some climate zones.

LOW-TECH AIR-CONDITIONING

Using a whole-house fan to flush hot air out of the house at night and replace it with cool outside air may be all the air-conditioning that's required. Closing windows during the day and relying on heat recovery ventilators to provide fresh air helps keep the house cool.

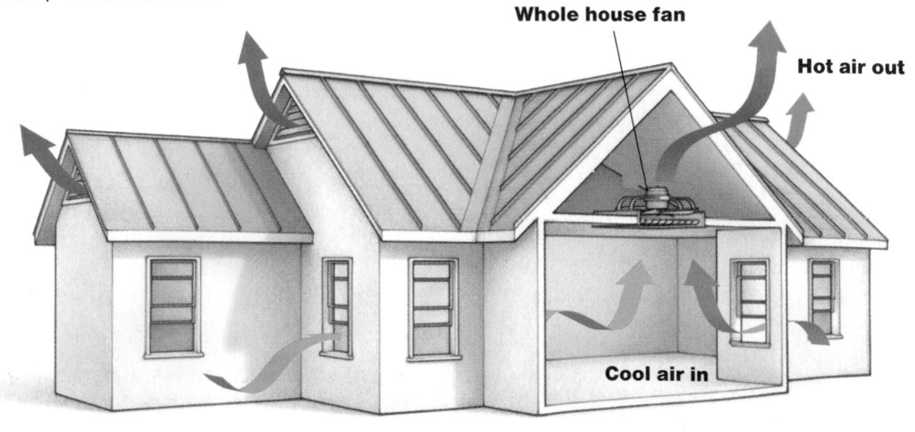

SEER RATINGS

Central air-conditioners, including air-source heat pumps, are rated by a seasonal energy efficiency ratio (SEER), which is the cooling output in Btu divided by the power input in watt hours over the course of a normal cooling season. The higher the SEER, the more efficiently the air-conditioner operates. Room air-conditioners use a similar scale, the energy efficiency ratio (EER), which is the ratio of cooling capacity in Btu and the input in watts at a given operating point.

conditioning is the norm, this simple approach may be all that's needed.

While passive measures can help, they do not necessarily eliminate the need for mechanical air-conditioning. Two-thirds of all houses in the United States have some kind of air-conditioning. As it is with choosing a heating system, the trick is finding a way to keep the house comfortable with the lowest possible energy cost. A good place to start is with the standard efficiency ratings for different kinds of air-conditioners.

Split central systems with ducts

In a central system connected to ductwork, a compressor and condenser coil are located outside and an evaporator inside. A refrigerant absorbs heat from inside air and is then pumped to the condenser. There, the refrigerant returns to a liquid state and gets rid of the heat—a heat pump running in reverse. Cool air is distributed through the house in ducts.

The capacity of air-conditioners is measured in tons, the equivalent of 12,000 Btu (this arcane benchmark refers to the amount of heat energy it would take to melt a ton of ice in 24 hours). Residential systems are typically sized between 1 and 5 tons, but in a net zero house, the loads are much smaller than average, and it is essential that the air-conditioner contractor use *Manual J* and *Manual D* to calculate loads and duct sizes, respectively, or a low-load computer modeling equivalent to get the system sized correctly.

Equipment that's too big uses a lot of energy and it cycles off and on frequently. These short bursts are not very effective in lowering humidity. Sealing ducts with mastic and locating ductwork inside the thermal envelope of the house are both important measures.

Government efficiency standards require a minimum SEER of 13, but models with a SEER of 23 are available—at a substantially higher cost.

SHORT RUN TIMES EQUALS HIGHER HUMIDITY

Oversize air-conditioning equipment is more expensive to install and operate than a smaller, but correctly sized, unit. And there's another problem: less effective humidity control.

New York City heat specialist Henry Gifford says it takes between 8 minutes and 12 minutes of run time for the condenser coil to remove excess humidity from the air. Shorter run times won't remove much moisture from the air, a chief benefit of air-conditioning in the first place.

Net zero houses have smaller cooling loads, making careful calculations for the capacity of air-conditioners essential.

Ductless mini-splits are easier to zone

Ductless systems use a common condenser/compressor outside to serve a number of individual evaporators inside, making the system much easier to zone. Although they're more expensive than standard central systems, they're ideal for houses in which heating ductwork doesn't exist.

An air handler can be up to 50 ft. or so away from the compressor/condenser, connected only by a small conduit for the refrigerant and power. Air handlers can be mounted on a wall or in the ceiling, and while they are more obvious than grilles for central systems, they don't make as much noise as a window unit. Mini-splits would be an appealing option when cooling an entire house would use more energy than occasionally high heat would warrant.

In a ductless mini-split system, a single outdoor compressor can serve air handlers in several rooms, making it easier to zone air-conditioning and heat and thus reduce energy consumption. Mini-splits are useful for retrofitting a house that doesn't have a duct system, but they also can be used in new construction.

In dry regions, consider a swamp cooler

An evaporative (or swamp) cooler uses less energy than standard air-conditioning equipment—and no refrigerant. They are a lower-cost option in areas with low humidity; because they add moisture to the air they are not well suited to damp environments.

Evaporative coolers work by blowing air over pads that have been dampened with water. Outside air is cooled down before it is distributed inside the house. High-capacity fans can move a lot of air, and the coolers don't require as much electricity to operate as refrigerated air-conditioner systems.

Ventilation for a Healthy House

Blower-door tests that help builders track down leaks in the building envelope are common fare these days. The tests, along

Air handlers in ductless mini-split systems are more conspicuous than built-in air registers, but they have several advantages over window-mounted air-conditioners. This wall-mounted unit is part of the Mr. Slim® system manufactured by Mitsubishi Electric Cooling and Heating Solutions.

ROOM-SIZE AIR-CONDITIONERS ARE LOW-COST OPTION

The lowest energy investment for keeping a single room cool is with a window-mounted air-conditioner. Such units can be somewhat noisy, and no fun to wrestle into a window opening, but energy efficiency has improved dramatically. Many run on a standard 115v circuit.

But in a zero energy house, insulation and air sealing should be so effective that minor blips on the heat and humidity scale aren't big enough to justify the installation of a room-size air-conditioner. The building itself should be a buffer to outside conditions that automatically demand a mechanical response in a conventionally built house. Moreover, using a room air-conditioner means the thermal barrier must be breached. After all, they're designed to go in after you open a window.

Running a window-mounted air-conditioner uses less energy than a central ducted system. But other options, such as flushing the house with cool air at night and keeping windows closed during the day, may be better options in a net zero house.

Evaporative coolers work by blowing outside air over pads that have been dampened with water, then distributing it through the house. They are well suited to dry climates, but not the humid South and East.

with careful air sealing, help eliminate air leaks that can rob a house of energy. While the industry benchmark for ventilation is 0.35 air changes per hour, zero energy builders aim for a fraction of that. They employ a variety of building strategies, which are covered in detail in chapter 2.

This is all well and good on the energy front, but tight houses are not especially healthy places without mechanical ventilation. Cooking odors, mold-inducing moisture, chemical emissions from building materials and furnishings, and pet dander are among the contaminants the house needs to exhale regularly.

This poses somewhat of a dilemma for the energy-efficient builder who has spent a lot of time and effort hoarding every possible kilowatt hour of energy. Taking all that conditioned air and dumping it outside means that incoming air must be heated or cooled, which takes more energy. The object should be to ventilate the house with the lowest possible energy penalty.

The general strategies are to push air out of the house and let makeup air in through vents or leaks in the building; to force air into the building and allowing stale air to find its way out through bathroom and kitchen vents or through leaks in the envelope; or to balance outgoing air with an equal amount of incoming air.

Of the three, balanced ventilation is more complicated and more expensive because it requires both supply and exhaust ducts to work properly. But it's also the most effective approach. There's a lot left to chance with ventilation systems that rely on only exhaust or supply fans, and there is a chance of drawing in contaminants from an attic or crawl space in the process. By maintaining neutral air pressure inside the house, a properly designed and installed balanced ventilation system isn't susceptible to those problems.

Capturing energy with a heat exchanger

To minimize energy losses, zero energy houses often rely on either an energy-recovery ventilator (ERV) or a heat-recovery ventilator (HRV). They are similar but not exactly the same.

In an HRV, incoming and outgoing air pass through a heat exchanger without mingling. In winter, as much as 80 percent of the heat in outgoing air is transferred to the fresh air pulled into the house. In summer, cooler air expelled from an air-conditioned house tempers the warm air drawn from the outside. HRVs are typically used in areas where heating is more important than cooling.

FRESH AIR FOR A TIGHT HOUSE

Superinsulated houses need a reliable source of fresh air. Heat-recovery and energy-recovery ventilators are designed to lower energy losses by passing incoming and outgoing streams of air through a heat exchanger. During the heating season, the temperature of cold air from the outside is moderated by outgoing stale air before the fresh air is distributed in the house. In summer, the process is reversed. The two airstreams do not mix.

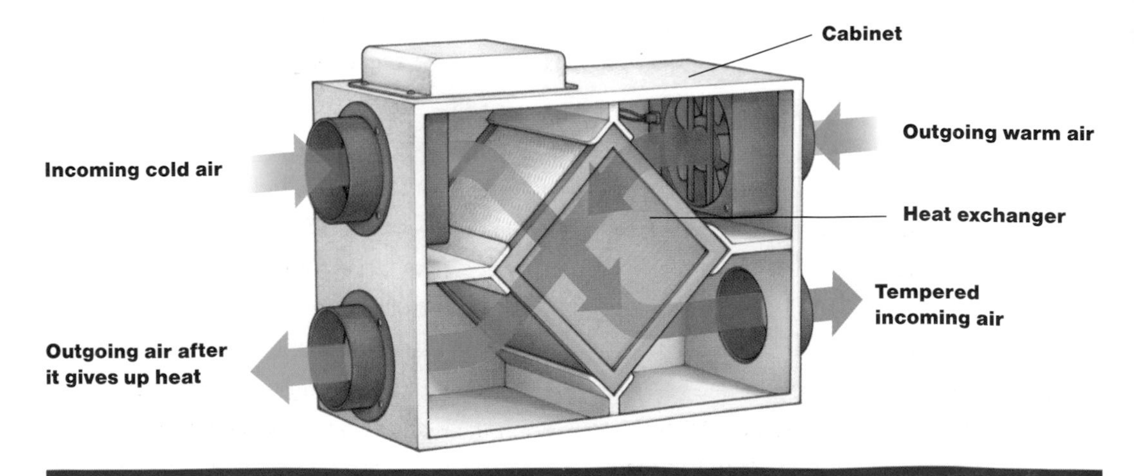

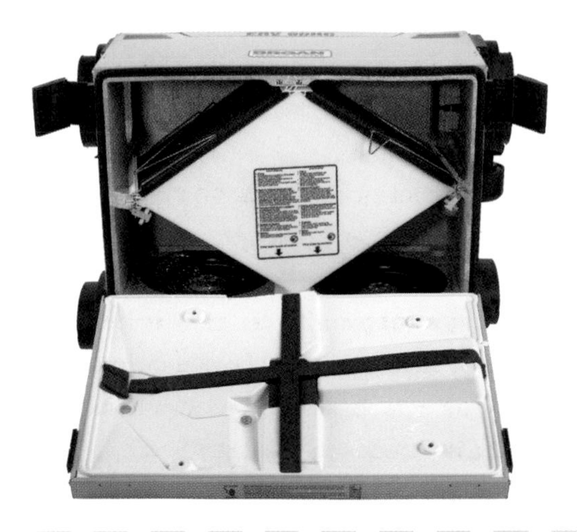

Whole-house ventilation is essential in a superinsulated, airtight house. A heat-recovery or energy-recovery ventilator can capture as much as 80 percent of the energy that would otherwise be lost in the process. Incoming and outgoing air pass through a heat exchanger but don't mix.

In an ERV, some of the moisture in the airstream also is transferred. This can be useful in a cooling climate. At least some of the moisture in incoming air is picked up by exhaust air, lowering demands on the cooling system and saving that much more energy. ERVs are most common in cooling climates, although they can also help maintain indoor humidity in cold climates during the winter.

Both ERVs and HRFs are usually built into a whole-house ventilation system. Ducting can become elaborate, and the devices add several thousand dollars in construction costs. But in houses tight enough for zero energy performance, they are an effective way of marrying energy performance and indoor air quality.

Heating Water

As is the case with every facet of energy use in a net zero house, the shortest path to conservation is often the simplest. It makes more sense to start with inexpensive solutions before investing thousands of dollars in the latest high-tech equipment.

How to keep hot water use to a minimum should by now be a familiar litany: low-flow showerheads and faucets, horizontal axis washing machines and energy-conserving dishwashers, plumbing layouts that keep plumbing runs short, insulated hot-water lines, and water circulation systems that eliminate the wait for water at the shower and sink. Anything that reduces the demand for hot water also reduces the amount of energy the house consumes.

In a net zero house, solar collectors (covered in detail in chapter 3) are the most logical way of heating water for domestic use. Depending on the solar potential at the site and how much hot water a family chooses to use, a solar collector can meet up to about 80 percent of hot-water needs. We just have to keep in mind that this is easier to accomplish in Tucson than it is in Seattle, and, in almost all cases, a solar collector will be connected to a backup heater.

Solar collectors are one of the most expensive water-heating options on the market, but incentives and tax credits can be significant, and the energy is completely clean

GETTING TO ZERO

Turn it down! Adjusting a thermostat by only 2°F (lower in the winter and higher in the summer) can mean a 4 percent reduction in energy use. On average, that small change also will keep 500 lb. of carbon dioxide out of the atmosphere every year.

and renewable. More important, it means that other renewable energy systems—a wind turbine or photovoltaic panels—can be that much smaller and less expensive because there is less imported energy to offset with site-produced energy.

Solar systems, however, are not a given, particularly when budget constraints collide with zero energy aspirations. At the Wheat Ridge net zero house (see "Affordable Housing Meets Zero Energy" on p. 233), project planners included solar hot-water collectors as well as photovoltaic panels. But the family that moved into the house used much less hot water than averages suggested. Increasing PV capacity to run a conventional hot-water tank would have been about $4,000 cheaper than installing the solar collector.

"The only conclusion that can be drawn from this simple analysis is that the conventional wisdom may not be true in all cases and additional investigation into the comparison of solar water heating and PV investments may be warranted," a report on the project said.

Coupled with other energy conservation measures and sensible water use, a high-efficiency electric heater won't necessarily break the net zero energy bank. Where hot-water use is relatively low, a gas-fired tankless heater (without a pilot light) may offer energy savings. Or in houses already equipped with heat pumps, adding a desuperheater can provide hot water.

A less common but appealing option is the heat pump water heater. It works just like an air-source heat pump, pulling heat from the air in the mechanical room and using a vapor compression cycle to concentrate and transfer the heat to a domestic hot-water tank. One manufacturer offers a unit that can be used to retrofit a conventional hot-water tank. AirGenerate™ claims its AirTap™ is 258 percent efficient at an

A tankless, or instantaneous, water heater is another option for domestic hot water. There are no standby losses, as there are with a conventional hot-water tank, but tankless units are combustion appliances that run on gas. Electric models are very heavy users of power. This one is installed in a zero energy house in Wheat Ridge, Colorado.

In an all-electric house with enough renewable energy capacity, a high-efficiency electric water heater may be the simplest way of providing domestic hot water.

ambient air temperature of 68°F and 166 percent efficient at temperatures as low as 32°F.

One plumbing addition that can be a big help regardless of the water heater type is drain water heat recovery. These are heat exchangers that transfer some of the heat from drain water to incoming cold water or to a dedicated storage tank. Nonstorage systems are simpler, consisting simply of a section of replacement drainpipe wrapped with copper tubing. But they are effective only when hot water is going down the drain at the same time that cold water is coming into the system (when someone is taking a shower, for example). One manufacturer claims energy savings can be significant—up to 40 percent—but comprehensive studies that document savings are few and far between.

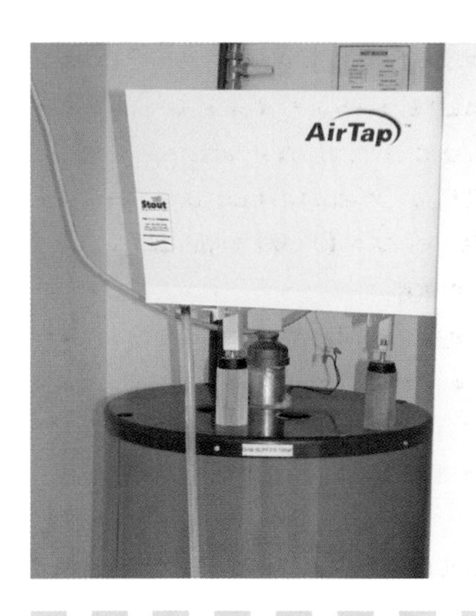

Heat pump water heaters are not widely used in the United States, but they are more efficient than conventional electric models. They tap into the latent heat of the air in the mechanical room, operating just like any air-source heat pump.

DRAIN WATER HEAT RECOVERY SAVES ENERGY

Hot water takes a lot of energy with it when it goes down the drain. In a drain water heat-recovery system, a heat exchanger made from copper tubing picks up heat that would otherwise be lost and returns it to the hot-water supply. Systems without a dedicated storage tank work only when hot water is being drawn at the same time it is discarded, as is the case when someone is taking a shower.

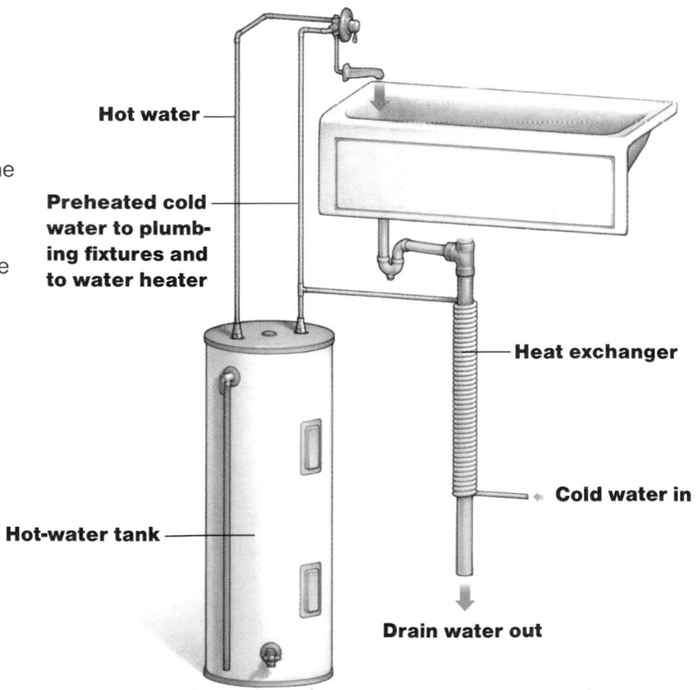

Case Study

A Solar Jewel in Boulder

A net zero energy house that's long on function and on form

While many net zero energy houses are long on function but short on form, the energy-efficient features of this house have been brought to life in a modern design that doesn't hide its technology. The lot that architect Michael Kracauer chose had an awkward shape with challenging access, but it had great solar potential.

SPECIFICATIONS

House size: 3,000+ sq. ft.

Average heating degree days: 5,577

Average cooling degree days: 736

Wall and roof construction: ICF basement; double 2×4 walls, 9 in. thick, 24 in. o.c.; 12-in. roof truss joists

Insulation type: 9¼ -in. Icynene® walls, 12-in. Icynene attic/roof

R-values: Foundation, R-30; walls, R-33; roof, R-43

Windows: Serious Materials double-Heat Mirror (R-14) with Duxton insulated fiberglass frames

Photovoltaic capacity: 7.2kW PV, should have enough excess power for at least one and possibly two plug-in cars; 140 evacuated-tube solar thermal for space heating and DHW

Heating source: Geothermal

Air-conditioning: None; closed-loop earth tube for cooling HVAC

Cost per square foot: $225

Designer: Michael Kracauer, architropic

General contractor: Morningstar Homes

With some of the strictest energy and green codes in the country, Boulder, Colorado, has a disproportionate number of net zero energy homes.

The back side of the house is oriented exactly south, providing ideal conditions for passive solar heating. The house is also outfitted with a 7.2kW photovoltaic system and 140 evacuated-tube solar thermal collectors for space heating and domestic hot water. As an architect, Kracauer wanted the solar systems to be architecturally integrated into the design of the house. The careful layout of the two active solar systems on their corresponding roofs, each with different angles for their different functions, and the sleek passive solar overhangs were possible because active and passive solar were both woven into the design concept from the beginning.

Taking advantage of the orientation of the lot for passive solar, the largest windows

The dual-pitch roof has a 7.2kW PV array for electrical needs and a 140-tube array of solar thermal evacuated tubes for space heating and domestic hot water.

and the main living spaces are located on the south side. Overhangs on the first and second floors help prevent the home from overheating. By contrast, the utility spaces and bathrooms are on the north side, with smaller windows.

Tightening Up the Envelope

As we've seen throughout this book, a zero energy home involves more than just solar technology; it also requires a tight building envelope. This home is built using double-staggered 2×4 stud walls with 9¼-in.-thick spray-in open-cell urethane foam insulation. The studs are placed at 24 in. o.c. (except in two small areas where exterior studs at 12 in. o.c. were structurally necessary). The two stud walls are separated by a 2¼-in. gap, which eliminates thermal bridging. This wall system uses just about the same amount of lumber as a 2×6 wall at 16 o.c. yet creates a uniquely thick wall system that allows additional insulation compared to a standard stud wall. The basement foundation walls are made from Greenblock® insulated concrete forms.

A double-wall system is a cost-effective way to reduce thermal bridging, allow more insulation, and control sound.

WHY WE LIKE ICFS

Insulated concrete forms (ICFs) made of polystyrene foam have a number of advantages over conventional construction materials. The top five mentioned by homeowners are:

- **Energy savings of 25 percent to 50 percent due to low air infiltration, high R-values, and high thermal mass.**
- **Reliable indoor comfort with virtually no cold spots or drafts.**
- **Exceptional noise reduction of up to 87 percent.**
- **Superior strength to withstand tornado- and hurricane-force winds up to 250 mph.**
- **Fire-resistant walls have up to a 4-hr. fire rating.**

In addition, ICFs are environmentally friendly: The concrete is an inert natural resource, and the ties, rebar, webs, and EPS foam can be recycled.

ICFs are nontoxic, have no air infiltration, and are very energy efficient. ICF construction currently makes up approximately 6 percent of new residential construction in the United States and is growing rapidly.

A generous amount of high-efficiency (U-0.07) glass brings in light and views while still keeping a tight envelope. The Duxton fiberglass frames have low conductivity and very low maintenance (and can be recycled at the end of their long life).

SeriousWindows

The house features windows and doors made with Serious Materials double-Heat Mirror glass (with an R-value of 14 at the center of the glass) and Duxton fiberglass frames insulated with spray foam filling. The fiberglass frames are more rigid than most other framing materials, which helps prevent the seals from stretching and deteriorating. Kracauer chose the same high R-value glass for the south side, rather than tuning the windows for the orientation, as is typically done. His energy modeling indicated that preventing heat loss by using the same glass throughout was more valuable than the extra solar gain. Insulated honeycomb blinds installed inside every window make a noticeable difference in heat loss.

Ventilation for a High-Performance Home

A heat-recovery ventilator (HRV) unit is designed to conserve energy by capturing the heat from stale air and preheating new fresh air coming into the building. The home is heated with NuAir's Enerboss® integrated HRV/air handler forced-air system, which uses hydronic heat from the solar thermal system. This system is designed for high-performance homes and is more efficient than using a separate HRV unit. Intermittently throughout the day, exhaust air for the HRV is ducted from the kitchen and bathrooms (both sources of warm, moist air), allowing the HRV to recover the energy in the warmed air from these areas. The exhaust fans are timed to turn on whether the rooms are being used or not. Lower fan settings are used for fresh air

CAREFUL AIR SEALING

Architect Michael Kracauer performed a blower-door test both before and after the drywall went up; the second test registered 0.058 NACH (or 0.77 at 50 pascals). As the energy rater put it, anything under 0.1 NACH is "off the charts."

Running a blower-door test before the drywall goes up is the smartest and most cost-effective thing you can do. It's a lot easier to fix any leaky area then rather than waiting till after the drywall is installed. Some air leaks were found around the windows and doors and also around the fireplace flue. These leaks were sealed with caulk and expansion foam.

when the bathroom fans turn on. The incoming replacement air into the HRV gets distributed throughout the house with the supply ductwork. The Enerboss unit integrates the HRV with the fan coil/air handler. By comparison, standard exhaust fans for bathrooms and kitchens create negative pressures in the home and exhaust energy from the house.

For kitchen cooking and odors, a recirculating fan built into the microwave oven exhausts the filtered air back into the room; a charcoal filter removes smoke quickly. This system was designed to avoid creating negative pressure in the house by simply exhausting air to the outside. Typically, the exhaust goes through the HRV, but the architect wanted to avoid kitchen smoke and odors contaminating the HRV filter.

An Enerboss HRV unit uses an advanced hot-water coil for maximum heat output. It has a variable-speed fan and a high-velocity option that is used when the exhaust fans are operating.

FORCED-AIR VS. RADIANT HEAT

In this house, a forced-air system made more sense than radiant heat for several reasons:

- **Forced air is less expensive.**
- **The Enerboss integrated air handler/ HRV provides much better ventilation distribution than a standard HRV.**
- **A radiant-floor system can sometimes overheat the home if too much thermal mass is used in the floors.**
- **An air-conditioning unit can be plugged into the existing ductwork if it is needed.**

When Is a System Too Efficient?

The solar thermal system had some design challenges. The evacuated tubes, which are about 7 ft. high by 7 ft. long, were so efficient that on hot days they overheated and produced steam. This caused undue stress on the system, and four water pumps were replaced over time. The solar installer eventually redesigned the system so that instead of a single continuous water pipe running through all the collectors, it is now split into two, with one pipe running through three panels and the other pipe through the remaining two panels. Previously, the water would run through all the tubes, and by the time it reached the last tubes it was steaming on the best solar days. By splitting the system in two, the water doesn't have a chance to get too hot. It also stresses the pumps less and has the added benefit of capturing the energy faster.

For this house, Kracauer maximized the passive solar gain by using a layout that takes full advantage of the great solar access, and he built a thermal envelope that functions like a thermos bottle. Solar thermal systems are, of course, very complex and have many variables, so they frequently require a lot of tweaking to get them working as efficiently as possible. Kracauer describes the PV system as "the easy plug-n-play part," compared to what needs to be done for solar hot-water heating in a net zero energy house. As he says, "Heating involves the entire design of the house, whereby one reduces the heating loads as much as possible so an affordably sized renewable energy system like solar thermal will be big enough." A solar-powered attic fan runs daily, and the attic is vented with a ridge vent and soffit vents. It is important to keep the attic well vented and cool in the

The domestic hot-water heater is made by Marathon, while the solar thermal storage tank was custom made with insulated spray foam. It holds 512 gal.

summer as attic spaces can become very hot during the intense Colorado summer, thereby increasing the building's cooling load.

All the Cooling the House Will Need

One of the house's most innovative features is the underground closed-loop earth tubes that circle the house, about 160 ft. each of 2-in. by 8-in. closed PVC pipe loops. The tubes are an extension of the return-air ductwork and are closed off by a damper until the thermostat calls for cooling. Then the warm indoor air is diverted through the tubes and reenters the home after being cooled naturally by the earth, which is at a constant 55°F. Unlike other earth tube systems (which use an open tube and are used to preheat or precool HRV replacement air), this system is a closed loop and is used only for cooling the building's supply air. A closed loop is less likely to have problems with condensation and therefore mold, and much more air can be pushed through the ducts than in an open-tube system, which is limited by being used only for replacement ventilation air through the HRV.

This system was inexpensive and, according to Kracauer, "since the house is so well insulated and will therefore need only a few degrees of cooling, this simple system should provide all the cooling the house will need." Kracauer's house is true net zero energy. It has yet to get through a full year through all four seasons, but from all indications it will perform as designed with energy to spare.

Metal roofing products are wind and fire resistant, durable, and provide a low-cost life cycle. In addition, cool metal roofing colors help reduce energy use in summer by increasing heat reflectivity.

Cooling is achieved by way of innovative closed-loop earth tubes that circle the house. The tubes are made from PVC and use the earth's constant temperature of 55°F.

CFL and LED lighting are used throughout the house. There is no gas line to the house, so all cooking is electric, including an energy-efficient magnetic induction cooktop.

Boulder, Colorado, has a number of net zero energy homes. New homes must be 50 percent more energy efficient than the 2006 International Energy Conservation Code (IECC).

Case Study

Reinventing the High Rise

Room-by-room thermostats and rigorous air sealing are the keys to energy efficiency and comfort in a New York City apartment building

Turning a multistory apartment building into a net zero energy consumer may be the Mt. Everest of residential design. When compared to a conventional single-family house, many apartment buildings don't have much insulation, and what's there isn't necessarily in the best place. Because of their height, the buildings induce a powerful stack effect that draws cold air in on lower floors as warm air is expelled through the roof. In winter, half the apartments feel like saunas, the other half are in the deep freeze. Moreover, there's not enough space on the roof for photovoltaic panels and solar hot-water collectors to serve everyone who lives there.

WHAT'S THE STACK EFFECT?

Inside a building with lots of air leaks, warm air rises to the top floor and escapes, to be replaced with cool air drawn in from below. This is not unlike what happens inside an industrial gas stack, hence the name for this phenomenon.

One obvious problem is a loss of heating efficiency. The cooler air being drawn in through leaks in the lower part of the building has to be brought up to a comfortable temperature as the air that's already been heated by the boiler wafts into the neighborhood. Leaky buildin[illegible] are also less comfortable because they're drafty. Replacement air could be pulled fro[illegible] just about anywhere, sometimes bringing c[illegible]taminants and moisture with it. Air leaks ar[illegible] doors and windows are common.

An effective air barrier, and especiall[illegible] sealing any penetrations at the top of the b[illegible]ing, helps control the problem. Heating and cooling costs are lower, and comfort is hig[illegible]

This apartment building in lower Manhattan looks ordinary enough, but it's really an energy miser. Thanks to air sealing, insulation, and carefully matched mechanical systems, it uses about 25 percent of the energy of a typical New York City apartment building.

It's not an inspiring tableau for net zero performance, and it may be unrealistic to think apartment buildings will ever compete on an energy footing with single-family houses, especially those with renewable energy sources. Even so, New York City architect Chris Benedict and her mechanical system designer Henry Gifford have found a way to reduce energy consumption for heat and hot water in buildings like this by as much as 75 percent when compared to existing New York City apartment stock–and to do so without increasing construction costs.

While Gifford rethinks the design of conventional mechanical systems, Benedict is convinced that an air barrier is the key to saving energy, improving comfort, and making her buildings more durable. "It's the most fundamental green thing that can be done in a building," she says.

Stopping the Hemorrhage of Warm Air

Many New York City apartment buildings are more than a century old and have very leaky building envelopes. Gifford, in fact, estimates that about half the heating load, and much of the cooling load, in a conventionally constructed apartment building comes from air leaks.

Benedict looks to create an air barrier not only for the building as a whole but also for each apartment. Techniques include the airtight drywall approach, sealing the outside of concrete block walls, and sealing every joint where one building material meets another. Blower-door tests at some of Benedict's apartment buildings are a very tight 1.2 ACH at 50 pascals of pressure (see chapter 1 for an explanation of air changes per hour rates). "So," she says, "the building now is like a bunch of very nice single-family homes all stacked on top of each other. The overall concept is to get solid material, seal all the joints, and get [each apartment] to perform as if it's its own little bubble."

Concrete block walls get a coating of Thoroseal®, a barrier to liquid water and also an air barrier. It does not block the passage of water vapor, but coating the block helps prevent air leaks that would reduce energy efficiency.

AIRTIGHT DRYWALL APPROACH

There are several ways to seal air leaks in a building, including the installation of polyethylene sheeting beneath the finish walls and ceiling. The plastic acts as both a vapor retarder and an air barrier, but it's very difficult to detail properly and it may actually trap moisture inside the walls, causing more problems than it solves.

The airtight drywall approach (ADA) is an alternative in which drywall itself, along with caulk and gaskets, is used to stop the flow of air. Joe Lstiburek, a well-known building scientist and consultant who now runs a Massachusetts company called Building Science Corporation, pioneered ADA in the 1980s.

Drywall is usually being installed anyway, so ADA is really a matter of taking a few extra steps to minimize air leaks rather than adding an extra building material (and one that is prone to damage during construction). Sealants and gaskets are flexible, so they can accommodate the normal movement of the building. These materials are typically applied where the drywall meets top and bottom wall plates and between the drywall and ceiling joists. In addition, a gasket is installed between the wall plates and the subfloor by the framers. There are a number of other details, including special electrical boxes, but they all are intended to stop (building scientists might be more precise by saying *retard*) the uncontrolled flow of air through the building.

Penetrations between apartments or between an apartment and a corridor or stairwell are sealed with an intumescent substance to produce a fire-rated assembly that's also an effective air barrier. This section of wall will ultimately be above the finished ceiling.

Insulation in conventionally built apartment houses is another sticking point. Many apartment buildings have structural walls of concrete block covered with brick veneer. Wall insulation is often placed in framed walls on the inside of the block. Benedict's approach is to place 3 in. of rigid mineral wool insulation on the outside of the block. It's attached with masonry ties and then covered with brick veneer. Now the insulation is a continuous layer, uninterrupted by the wires, pipes, and ducts found inside interior walls. The R-30 roof is insulated with 6 in. of extruded polystyrene board.

Keeping the mass of the block wall on the warm side of the thermal barrier also is an advantage, as researchers from the Florida Solar Research Center learned with a prototypical near-zero house in Lakeland, Florida (see "High Heat" on p. 42).

Smaller Heating Loads, Smaller Boilers

Gifford's forte is heating systems (his surprisingly well attended tour of New York City boilers was the subject of a 2003 profile in *The New Yorker* magazine), and he begins with Benedict's detailed calculations for heat-

The brick façade conceals a 3-in.-thick layer of mineral wool insulation. Keeping the mass of the concrete-block structural wall on the warm side of the building helps even out spikes in indoor temperatures.

The ½-in.-thick plastic mesh behind the brick keeps the inside of the building drier. Concrete pushed toward the inside as bricks are laid can't form a dam that would block weep holes at the bottom of the wall.

ing and cooling loads. Given the performance of the buildings that Benedict has designed, he's able to specify smaller than usual boilers along with smaller pumps to move hot water around the building.

For example, the boiler that Gifford chose for one of Benedict's high-performance apartment buildings at 228 East Third Street in lower Manhattan is a Weil-McLain®580 with a natural gas input of about 639,000 Btu per hour. It supplies both heat and hot water for the 22 apartments in the building. By comparison, a boiler in a conventional single-family home could easily be rated at 100,000 Btu per hour. Zone pumps to move hot water around the building are correspondingly smaller than normal, rated at 1/2 hp when normally they would be between 3 hp and 5 hp.

Most important, each room in each apartment gets its own thermostat for precise temperature control. The thermostatic radiator valves are capable of regulating hot water from full flow to a trickle, thereby helping solve the overheating problem common to many apartments. There is little variation in temperature between apartments on top and bottom floors, a rarity for this type of building.

On top of a conventional torch-down roof are two layers of rigid foam insulation over a membrane that acts as a capillary break. This approach keeps the thermal mass of the roof on the warm side of the insulation. The foam is capped with landscape filter fabric and stone.

Ventilation System Uses Building's Shell

Just as in single-family construction, airtight apartment houses make mechanical ventilation a necessity, but the size and height of a multi-family apartment building is a complication. According to Gifford, apartment buildings typically rely on a buildingwide ventilation system that pulls stale air out of each apartment via bathroom and kitchen vents and ducts it to a common exhaust at the top of the building. Apartments on upper floors have pretty good ventilation, but performance falls off on lower floors and at the bottom of the building barely any air moves at all.

Benedict's neat solution is to take advantage of holes inside the precast concrete planks that are used for floors and ceilings. Running the full 36-ft. length of each plank are 5-in.-dia. holes, which are there to reduce the weight. With the addition of Panasonic® in-line fans, these tubes become ventilation ducts. Air is pulled from kitchens and bathrooms, forced into one of the holes and then out of the building through an exhaust grille on an exterior wall. Small vents in each bedroom window introduce a trickle of fresh air into each apartment.

Installation is simple because there is very little ductwork, and each apartment has its own continuously operating ventilation system that draws about 25 percent of the electricity of a conventional system.

Stale inside air is pulled through cavities inside the concrete planks that form the walls and ceilings and is expelled through the side of the building. This exhaust line will be covered by a grille in the finished brick exterior (left).

Combined, all of these steps reduce energy consumption for heat and hot water to about 25 percent of what it would be in an average New York City apartment building. Although improved energy performance could come with such upgrades as more insulation and better windows, Benedict and Gifford have gotten this far without increasing overall construction costs.

A NEW TWIST ON HIGH-RISE VENTILATION

New York City architect Chris Benedict devised a more effective approach to ventilating apartment buildings by taking advantage of air cavities inside the cast-concrete planks used for floors and ceilings. Small fans draw air into the passages and expel it through a grille on the face of the building. Air is admitted at small vents at windows in each apartment.

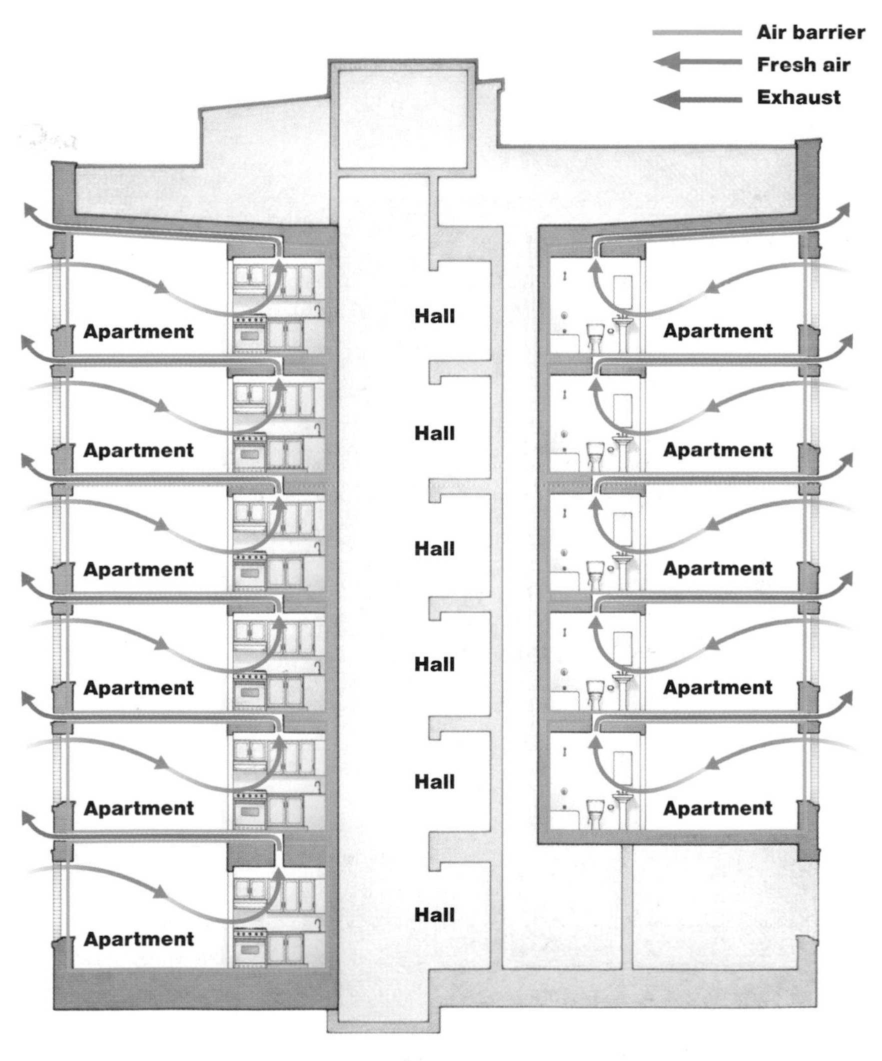

Case Study

All-Electric House: $1.16 a Day

An affordable design for Habitat for Humanity with a flexible floor plan and very low energy costs

No single designer can take credit for this near zero building, the fifth and largest low-energy house in a Lenoir City, Tennessee, Habitat for Humanity neighborhood. By the time the house was built in 2005, the Oak Ridge National Laboratory had collaborated with the Department of Energy's Building America Program, the Tennessee Valley Authority, and Habitat for Humanity on four earlier versions. ZEH-5, as researchers called it, summed up what they had learned.

SPECIFICATIONS

House size: 2,632 sq. ft.

Heating degree days: 3,200

Cooling degree days: 1,300

Wall and roof construction: Structural insulated panels

Insulation type: Expanded polystyrene

R-values: Foundation walls and slab, R-10; walls, R-21; roof, R-35

Windows: Double-glazed, low-e, with vinyl frames; U-0.34, solar heat gain coefficient 0.33

Photovoltaic capacity: 2.2kW

Heating and air-conditioning: Ground-source heat pump

Cost per square foot: $85

Designer: Oak Ridge National Laboratory, Building America

General contractor: Habitat for Humanity

FROM WARTIME ORIGINS, LAB'S INFLUENCE GROWS

The Building Technology Center that helped develop the near-zero energy houses in Lenoir City is part of an enormous research institution with interests in everything from molecular biology and holography to metals processing and mouse genetics.

The Oak Ridge National Laboratory near Knoxville, Tennessee, was created in 1943 as part of the Manhattan Project, the country's top-secret effort to develop an atomic bomb. It later became a center for the study of nuclear energy, but with the creation of the U.S. Department of Energy in the 1970s its mission was broadened considerably. In all, Oak Ridge now has more than 4,300 employees and a budget of $1.4 billion. It's the Energy Department's largest science and energy lab.

Just about anything having to do with energy-efficient building techniques and materials is fair game for scientists in the building technologies program. Research and testing covers wall and roof assemblies, foundations, insulation, and other materials that affect building performance.

All of it is a gold mine for anyone looking for information on energy-efficient building practices. On-line calculators, research articles on many building topics and R-value test results on structural insulated panels (SIPs), wood-frame construction, and steel framing are among the many resources the lab offers for free through its Building Envelopes Program.

This four-bedroom, 2,600-sq.-ft. Habitat for Humanity house in Lenoir City, Tennessee, was designed with the help of the Oak Ridge National Laboratory to use substantially less energy than a conventional, code-compliant house in the same area. Electricity costs are a little more than $1 per day.

With an insulated, walk-out basement, ZEH-5 is really a two-story house of more than 2,600 sq. ft. That's too big for Habitat for Humanity guidelines, so researchers essentially divided the building into two parts. While researchers measured energy consumption in the year after the house was built, the upper floor was used as an office and the lower floor was essentially an insulated crawlspace. Later, the two halves of the building were joined with a set of stairs, and the house turned over to its new owners.

Designers added an extra 2 in. of extruded polystyrene insulation to the 8-in.-thick roof panels for a total R-value of 35. Seams in the panels were sealed with caulk, foam, or a peel-and-stick membrane.

Airtight Shell Made from Structural Insulated Panels

Both the walls and roof of the two-story house are structural insulated panels (SIPs). By zero

Structural insulated panels with cores of expanded polystyrene foam form both the walls and the roof of the house. Walls are rated at R-21, somewhat low by zero energy norms, but thorough air sealing helps boost efficiency.

energy standards, the house is by no means overinsulated. Wall panels, which are 6½ in. thick, have a core of expanded polystyrene (EPS) foam and are rated at R-21. The 8-in.-thick roof panels, also with EPS cores, are capped with an additional 2 in. of extruded polystyrene insulation for a total R-value of 35. The walk-out lower level has R-10 exterior insulated-block walls and a concrete slab insulated with extruded polystyrene board, also R-10.

Builders sealed all joints between the insulated panels with caulk, foam, or peel-and-stick tape. As a result, the house has a low rate of air leakage—about 1.7 air changes per

THE VALUE OF AN INSULATED BASEMENT

An insulated foundation helps make the near zero house in Lenoir City, Tennessee, far more energy efficient than a conventionally built home in the same area. Rigid insulation on the outside of the concrete block foundation wall, rated at R-10, also serves as a buffer for any movement in surrounding soil.

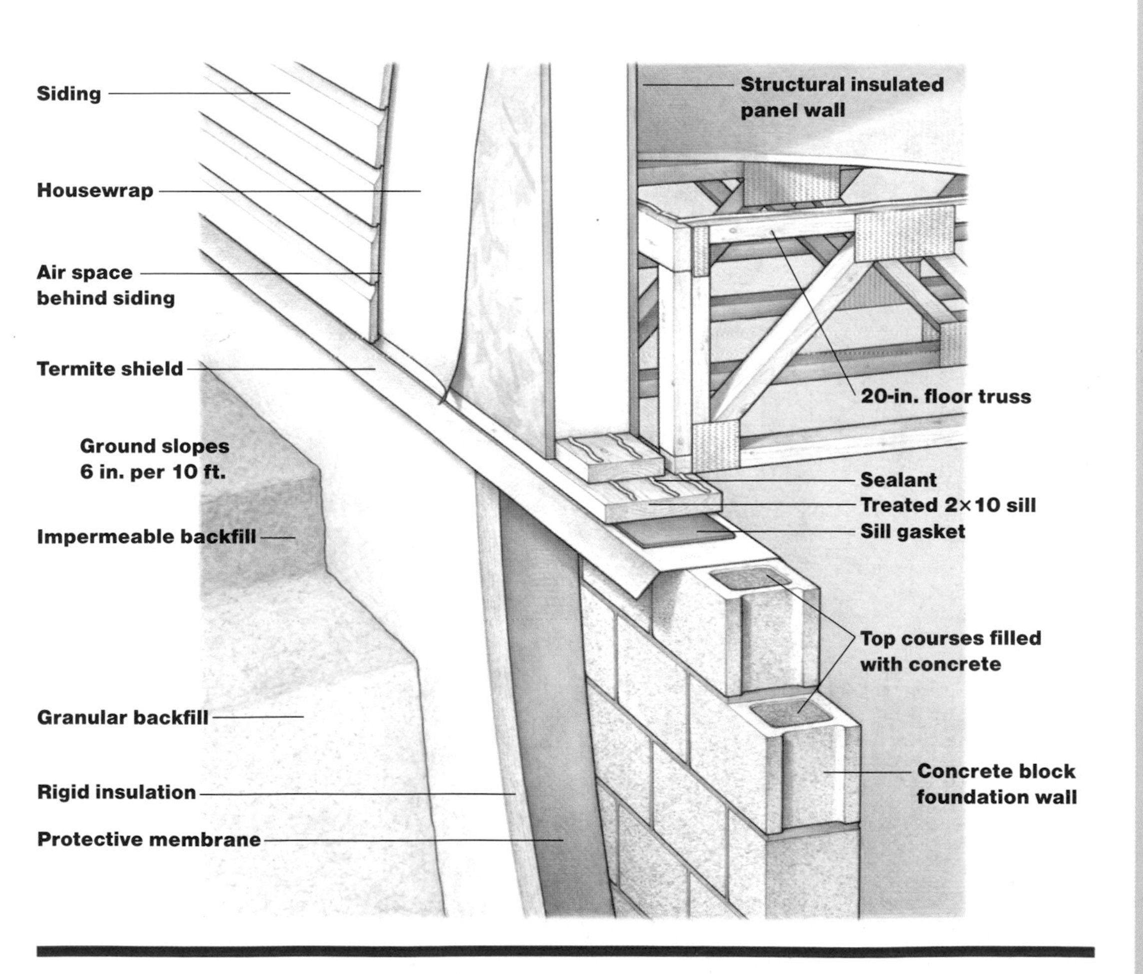

hour (ACH) under 50 pascals of pressure—roughly one fifth the rate in a comparison benchmark house. For ventilation, outside air is ducted to the supply return side of the air handler and controlled by a motorized damper control. A bathroom exhaust fan is wired into the system to help balance air pressure inside the house. The system is designed to provide about 64 cu. ft. of fresh air per minute, the amount recommended by the American Society of Heating, Refrigerating, and Air Conditioning Engineers (ASHRAE) for a house of this size.

Although the house is utilitarian by U.S. standards, designers were able to give the floor plan a measure of future flexibility. A site-built ridge beam runs the length of the house, so with the exception of a 12-in.-thick utility wall between the kitchen and a bathroom/laundry, none of the interior partitions is structural. Bedroom walls could be moved to accommodate a new floor plan without disturbing the basic framework of the house.

Low-Impact Geothermal Heat

Heating and cooling come from a 2-ton ground-source heat pump. That's not exactly unusual, but designers found a way to install the 1,500 ft. of tubing for the system in a way

LESSONS LEARNED

The walls and roof of this house are made from structural insulated panels, or SIPs, which combine a core of insulating foam with outer layers of oriented strand board (OSB). Here are10 tips from the research team that designed and built the house on working with insulated panels in a zero energy design:

- **Train yourself about zero energy construction or get trained people involved in the project as early as possible.**
- **Protect SIPs from damage. Stack them on site dry and flat and off the ground, and organize them so they go up logically—that is, panels on top of the stacks go up first.**
- **Get the right equipment for rigging and lifting heavy or awkward building components. A boom truck, for example, may be needed for heavy lifting, and specialized tools required for cutting panels where necessary.**
- **Make sure the foundation is accurate.**
- **Provide a drainage plane on both walls and roof to protect the panels from water damage.**
- **Minimizing air leaks is a primary goal, so know the details of connecting panels and running wires.**
- **Check drawings for accuracy and make sure that details on allowable spans are correct. Verify that point loads are adequately supported.**
- **Attach solar panels to the roof to minimize penetrations of the roof membrane. Here, clips for the panels were fastened to raised metal seams without disrupting the roof membrane and increasing the risk of leaks.**
- **A single ridge beam to support the roof panels makes construction faster. Plan the distribution of labor and arrange for a boom truck to get the job done quickly and safely.**
- **Devise a sound construction plan to match the resources that will be available, and make sure that all subcontractors understand what it is.**

Rigid fiberglass insulation applied over the concrete block foundation served two purposes. In addition to providing thermal insulation, the board helped isolate the foundation from any freeze-thaw movement in the soil caused by buried geothermal tubing.

that holds promise for houses on cramped suburban and infill lots.

Geothermal systems with tubing placed in the ground horizontally typically require sizable building lots because trenches take up so much room. Drilling vertical wells for a closed-loop geothermal system is an alternative, but it's also more expensive and well out of the price range of affordable housing. At ZEH-5, the high-density polyethylene tubing was buried in a total of 244 ft. of trench that required no more excavation than what the foundation and buried utilities already provided. At the time, the ground loop was experimental and not certified by the International Ground Source Heat Pump Association. But it worked nonetheless and the concept was later used for two 3,000-sq.-ft. houses built near the Oak Ridge National Laboratory.

Using common utility trenches for some of the tubing saves money and space, but it requires careful planning. Trenches must be deeper than usual and water lines and geothermal tubing have to be installed not only in the right sequence but also far enough apart to prevent the water supply from freezing during the winter as the geothermal system draws heat from the soil.

Geothermal tubing is usually kept at least 10 ft. away from the foundation wall so the earth around the foundation won't freeze. But in this climate, designers believed that foundation insulation plus perimeter drains would reduce the risk. According to Jeffrey E. Christian, who directs the Building Technology Center at Oak Ridge, even if the ground should freeze, the fiberglass insulation board on the foundation exterior would act as a "slip plane and compression cushion" to prevent any damage.

Photovoltaics and Solar Hot Water

ZEH-5 has two solar collectors totaling 40 sq. ft. on the roof for domestic hot water and an 11w PV panel to power the hot-water pump. Initially, the house did not have its own photovoltaic array for generating electricity. Designers borrowed data on electrical generation from a 2.2kW PV system that had been installed right across the street, at ZEH-4. Orientation toward the sun and the roof angle are identical, so Christian was confident the numbers could be extrapolated accurately for the newer house. The $18,500 PV system generates about 8kWh per day, or roughly 31 percent of the total electrical consumption.

ZEH-5 got its own PV array in June 2008, a dozen 208w panels from Sharp that produce about 2.5kW at peak, a slight increase over the system on ZEH-4. Panels are attached with clips anchored to the standing seams on the roof, a neat system that doesn't require any through-the-roof penetrations.

Although the PV system leaves the house in the near zero energy category, its daily electrical consumption of $1.16 per day is an attractive alternative to the $5.14 a day the benchmark comparison house would need. As important, solar-generated electricity reduces peak summertime loads on the utility by 4kW, which is worth, according to Christian, about $2,000 in avoided costs for the utility.

Jeff Christian, director of the Building Technology Center at the Oak Ridge National Laboratory, with the inverter that converts direct current generated by photovoltaic panels into the alternating current used in the house. Houses in the Lenoir City community have net meters to measure the amount of electricity exported to the utility's grid plus a meter showing output of solar panels.

Solar hot-water collectors and photovoltaic panels take the sting out of utility bills. Here, a small PV panel provides power for the circulation pump on the hot-water collector.

5

Living a Net Zero Life

Jane Bindley, a botany major in college who later went on to work as a physical therapist, considered herself in tune with the physical world around her. At least she did until one evening in 2004 when she attended a lecture just down the street at the Appalachian Mountain Club in Boston.

Ross Gelbspan, a Pulitzer Prize-winning reporter, was the guest speaker. Gelbspan had worked at three big metro dailies, *The Philadelphia Bulletin, The Washington Post,* and *The Boston Globe,* and later published two widely read books on climate change, *The Heat Is On* and *Boiling Point.* When he finished speaking that night about the havoc that global warming was serving up, Jane Bindley's world had been turned upside down.

"It was horrifying," Bindley said in the summer of 2008 in the living room of her lakeside New Hampshire home. "I had considered myself to be very politically astute. I had worked on lots of campaigns. I'd always been aware of the environment. I was just knocked out."

Bindley e-mailed Gelbspan and asked for a copy of his speech, and he promptly sent her the 36-page text. She got interested enough in spreading the word that she became a volunteer lecturer with the Green House Network. She also audited classes at Harvard Extension School on climate change and sustainability. In one class, the assignment was to find something in her immediate environment that could be changed—getting a local coffee shop to switch from plastic foam cups to paper cups, for example. "It was like turning on a light switch," she said. "I started looking at all the opportunities there were in life to make changes."

Bindley worked locally, convincing the organizing committee for her college's

Jane Bindley turned a 1970s ranch into a net zero energy home, partly in hopes that it could become an educational tool for others interested in buildings with a low environmental impact.

50th reunion to make it a green event. She launched a climate change awareness group in Weston, Massachusetts. By the time she began looking for a vacation house in New Hampshire, the grandmother of 11 knew enough about green building and sustainability in general that she insisted on turning the tired 1970s ranch on the edge of Squam Lake into a net zero remodel (see "High-End Lakeside Retrofit" on p. 154).

She got a new way of looking at her life, not just a newly rebuilt house, and she was eager to share its lessons with anyone who was interested.

David Pill, commercial architect turned low-energy advocate, at home in Charlotte, Vermont: "I realized that I was making a negative contribution by doing all this work. I wondered, 'How can I make less of an impact?' "

No light bulb went off suddenly for David Pill, at the time a Massachusetts architect with a brisk business in commercial design. Although he had a few residential clients, Pill had his hands full designing a series of retail stores for Verizon and handling other commercial work that job steered his way.

But by the early 1990s, Pill was starting to get concerned about the environment. "I realized that I was making a negative contribution by doing all this work," he said. "I wondered how I could make less of an impact, so I just started doing a lot of research on sustainable design."

He joined the Northeast Sustainable Energy Association and began attending workshops and the group's annual conference in Boston. He stockpiled information on the materials and building techniques that would allow him to design environmentally friendly buildings. By then, Pill and his wife, Hillary Maharam, had moved to Winchester, Massachusetts, and had launched their own architectural firm. They were living in a 1711 house where, Maharam says, "colonies of mice" ran freely through the walls.

"The house burned oil like crazy," Pill said. "We were just about to do the second phase of the renovation, which would have been totally reinsulating it and tightening it up, but we decided we wanted a change of lifestyle, so we moved. I never had this vision that we were going to live in a zero energy house. It wasn't like that. The vision was we wanted to move out of this area and have a more sustainable lifestyle where we could grow food and have animals and just be in a nice landscape. So that was our impetus."

Yet somehow their move to Vermont and the subsequent construction of a new house became something more than that. The family of four got the rural setting they wanted, a hilltop site and views of a rolling, tree-covered landscape dotted with farms, space for vegetable gardens, an old apple orchard, and a spot for a chicken coop. David also got the itch to build a house so energy efficient that it would be a net producer, not consumer, of energy (see "The Good Life in Vermont" on p. 138).

It's Not Just Low Energy Bills

On its most pragmatic level, net zero building is about the elimination of energy bills—from the gas company, the local electric utility, the oil company. There must be a special glee in opening a statement from the power company and finding a credit instead of a debit or in strolling up to the two-way meter and watching the kilowatt hours of electricity flow *into* the grid. The backdrop for life in the Pill-Maharam household is the *whuh, whuh, whuh* of the wind turbine blades spinning atop a tower in the field below the house. To Pill, the turbine is an "energy-producing kinetic sculpture," but it's also a reminder that they are energy self-sufficient.

David Pill and Hillary Maharam tend the vegetable garden outside their net zero home. Moving to Vermont and building the house was part of their plan for a more sustainable lifestyle.

A 10kW wind turbine in the field below the Pill-Maharam house provides all the energy the family of four needs.

Spending less on energy is attractive, but there's more to it than that. Bindley, Pill, Maharam, and others who have made the investment in net zero houses also seem to appreciate that the energy choices they make have a ripple effect on their world in general. Green builders call this *sustainability,* but a generation or two ago we might have called it thrift or prudence or husbandry.

"It's something about self-sufficiency and being on this property and creating this homestead," Pill said last summer as he sat across the dining table from Maharam and talked about their experience with the house. "That's a piece of it. The other piece was if we're building a new house with all new materials, or a lot of new materials, and making that kind of impact, I felt it was really important to at least not make an energy impact."

Trained as an architect, Hillary Maharam later left the field and worked as a landscape designer and freelance editor. One objective these days is to learn how to grow and preserve food.

Pill and Maharam are learning how to grow their own food. "That's my job now," Maharam said. By late June, they were eating out of their garden and talking about raising chickens not only for eggs but also for meat. Maharam envisioned basement shelves sagging under the weight of canned tomatoes and beans. Pill has been restoring the apple orchard on the property.

Bindley has planted a garden, too, which is irrigated with water stored in a 1,000-gal. cistern. Rain gutters on the house keep the cistern full, and photovoltaic panels on the roof power the pump for the drip irrigation system. She has a lawn, a species of drought-resistant fescue that shouldn't need much attention. In front of the house by the lake, the grass is growing around a slab of granite that serves as an outdoor coffee table. The lawn is pleasantly unkempt.

Gaining a New Appreciation for Energy

Sinking a lot of money into photovoltaic panels or a wind turbine has a way of making people acutely aware of how much electricity they're using. By definition, net zero houses produce as much power as they consume on an annual basis. Sometimes the renewable energy systems are producing more than the house needs and the grid gets what's left over, and sometimes the systems can't keep up with demand. But at the end of the year, the ledger must show a net gain, and that doesn't happen without building a very tight house to lower energy demands and then managing electrical consumption carefully.

It's feasible, of course, to overload a house with electricity-producing equipment so that in the end there's no question it can reach that magic net-zero plateau and get the bragging rights that go with it. But the opposite seems true on the net zero and near net zero houses we found. Production and consumption were carefully calculated from the start so that no more capacity was installed than what

A vegetated roof caps the car-tow garage behind the Bindley house. In addition to being a net zero energy consumer, the home also won Leadership in Energy and Environmental Design Platinum certification.

Jane Bindley picked a type of fescue grass that needs little watering or maintenance. The yard is never cropped to within an inch of its life.

The vegetable garden outside Jane Bindley's New Hampshire home gets its water from roof gutters, via an underground cistern and a drip irrigation system powered by photovoltaic panels.

designers thought homeowners would need. That's in keeping with the green philosophy of using no more resources than necessary, but it also reflects the high cost of solar hot-water collectors, photovoltaic systems, and wind turbines. None of the energy produced by these systems can really be considered cheap.

So the owners of net zero houses start looking at every possible energy leak, the unexpected places where electricity needlessly dribbles away. Bindley can tell you exactly when her photovoltaic system went online, April 22, 2008 (Earth Day), and how much surplus generation the house could claim by April 22 of the following year. She can also tell you about the phantom loads that Marc Rosenbaum, her energy designer, tracked down to eliminate waste. There was the washing machine that drew 15w even when it wasn't running, the irrigation pump controls that got between 5w and 7w, and the garage door opener that needed 7w to remain in standby mode.

All of this combined is barely enough to keep a single light bulb burning, and not a very bright one at that, but it was galling to Bindley that there was any waste at all. During the day, she doesn't need lights inside the house. She unplugs the washer when it's not in use, and she's about to tackle the garage door problem by disarming the opener completely and opening and closing the doors the old-fashioned way. "I'm going to lengthen the [pull] rope so I don't have to use electricity to open and shut them," she said. "That's just a no-brainer."

At the Pill-Maharam household, there is but one incandescent light bulb in the house, a fixture in the stairwell leading to the basement. They bought it just because they liked it, but Pill is quick to point out that it's rarely on. Everything else is fluorescent. During the day, the house interior is well lit even on a cloudy day, thanks to a design that kept the

Even on an overcast day, there's no need for electric lights inside. Jane Bindley's desk faces Lake Squam.

house relatively narrow and open and its long axis facing south. They found a dishwasher that could be completely turned off when not in use, bought an efficient induction cooktop, and connected electronic devices to power strips that could be switched off to eliminate the trickle charges.

The couple and their two children have a minimalist style to begin with, so you won't find a lot of gizmos and gadgets plugged in everywhere. "I have an awareness of every single thing that's plugged in anywhere in the house," Pill said. "Even the toaster oven has a little switch in the outlet because that has a little phantom load, so that goes off after we're done. And the computer is on a power strip that just gets turned off. We're not using energy when we don't need to be."

Pill keeps detailed records of the electricity produced by their 10kW turbine. A separate meter in the basement next to the inverter records total electrical output, and a meter set on a post near the house records the amount of electricity that's imported and exported. A digital screen switches automatically from one mode to another: consumption, production, consumption, production–24 hours a day, 7 days a week. Record keeping without end.

Buying high-efficiency appliances, using fluorescent or LED light fixtures, and eliminating phantom electrical loads are all important strategies for reducing energy consumption. If careful design and construction produce a superior building envelope, the lifestyle choices of its occupants become even more significant than they would normally be.

David Pill explains the operating of a digital electric meter outside his Vermont home. It alternates readings of incoming and outgoing electricity.

Mechanical Systems Are More Complicated

Renewable energy systems can be foreign to those of us whose previous interaction with energy delivery amounted to adjusting a thermostat on the wall or phoning the propane dealer when the tank got low. Although photovoltaic panels have no moving parts and should last 25 years or more with virtually no maintenance, inverters and meters add another layer of complexity. Any system with a battery bank will need minding. Active solar hot-water systems introduce more valves, controllers, and pumps to the mechanical room, all of which must be checked and, occasionally, repaired or maintained.

That's the reality, but even mechanical novices like Bindley are willing to learn. With two mechanical rooms in the basement—one for three hot-water tanks and circulation controllers and another for the geothermal heat pump, inverter, and main electric panels—there's a lot of stuff to keep track of. But standing in a thicket of pipes, gauges, electronic boxes, and valves, Bindley can explain what most of the equipment around here is supposed to do.

She'll need some help flipping the right breaker to power down the heat pump for the season, and a few of the electronic controllers are still mysterious. But she has all of the fundamentals and easily explains a somewhat complex solar hot-water collection and storage system, the geothermal heat, and a variety of other components.

Pill and Maharam also have a lot of equipment to keep track of, and their experience is complicated by a renewable energy source that needs periodic maintenance.

A daunting network of pipes, tanks, valves, and gauges fills one of two mechanical rooms at Jane Bindley's New Hampshire house. With a few exceptions, she knows what all the equipment does.

A gin pole attached to the base of the turbine tower at the Pill-Maharam house is used to lower the tower for maintenance, a chore required every couple years.

Although the wind generator has been trouble-free once a balky inverter was replaced early on, the turbine will need routine maintenance. To make that possible, Pill chose a type of tower that pivots at the base and can be lowered to the ground with the help of a winch and gin pole. It's not a job that Pill is anxious to do himself. Instead, he'll invite the installer back.

Barring outright equipment failures, maintenance should be manageable. But there is more equipment to keep running, increasing the amount of time it takes to keep tabs on mechanical systems and increasing the chance that something will go wrong and need repairs. Photovoltaic panels should need the least amount of maintenance; wind turbines, the most.

Fighting the Perceptions of High Cost

The initial cost of renewable energy systems is high, even with new federal tax credits that knock 30 percent off the cost and state incentives that further increase potential savings. Introducing heat-recovery ventilators and substantially bumping up the levels of insulation in net zero buildings also add to the cost of construction. All of this understandably fuels the perception that net zero houses are for the well-to-do and not the rest of us.

Some projects bear that out. But operating costs are lower, and several projects explored in this book prove that net zero building doesn't have to be hugely expensive. Because they were constructed under the Habitat for Humanity program, the houses in Wheat Ridge, Colorado (see "Affordable Housing Meets Zero Energy" on p. 233), and Lenoir City, Tennessee (see "All-Electric House: $1.16 a Day" on p. 210), were modest in cost–$116 per square foot in Wheat Ridge and $85 in Tennessee.

One way of keeping costs down is to use conventional building materials and techniques, which is what makes the Wheat Ridge house especially appealing. Exterior walls are double 2×4 construction–two 2×4 framed walls with a space in between for additional insulation. The insulation itself is fiberglass batts, not very expensive and requiring no special expertise to install. Many builders abandoned 2×4 construction in cold climates years ago because walls didn't allow enough insulation, and fiberglass batts are prone to air leaks that lower thermal efficiency. Yet the project worked, suggesting that zero energy buildings could be made available to many more homeowners of average means if only consumers knew to ask for them and builders took a few extra steps during construction.

Net zero houses, of course, can cost a lot more than the Wheat Ridge house. Pill's Vermont house cost about $200 a square foot to build, and at 2,800 sq. ft. it's hard to argue the project is a template for affordable housing. Pill, however, says some of the cost can be attributed to interior finishes and materials he and Maharam chose: expensive door hardware, plaster walls instead of drywall, more time-consuming detailing. He thinks the same basic house, without some of those finishing touches could be built for between $140 and $150 per square foot. That's still

Using conventional building materials and techniques helps keep costs down, as in this Habitat for Humanity house in Wheat Ridge, Colorado. Double 2×4 exterior walls allow space for extra insulation.

not inexpensive, but it's a lot less than many comparable custom homes with very average energy performance.

An obvious point worth considering is the impact that house size makes on overall costs. Houses have gradually been getting bigger, growing from an average of 1,660 sq. ft. in 1973 to more than 2,400 sq. ft. in 2005. Zero energy designers, borrowing a page from the green building manifesto, can lower costs significantly by making houses only as big as they need to be. If, for example, a family could make do with 2,000 sq. ft. and assuming Pill's estimates are correct, a net zero house in a cold climate could be built for $280,000. That's $19,000 less than the price of an average house built in 2008, according to the U.S. Census Bureau.

Further, energy costs are predictable from the moment the house goes on-line because the cost of its renewable energy systems is rolled into the mortgage. Suppose a house includes a 7.5kW photovoltaic system (that's pretty big) priced at $8 per installed watt, a $60,000 system. After the federal tax credit, the cost has dropped to $42,000, which would cost $252 per month over the life of a 30-year mortgage at 6 percent. Any state incentives would only sweeten the deal.

The cost of heating and cooling this hypothetical net zero house won't change. Compare that to a home heated with, let's say, fuel oil. To use the pricing ups and downs of 2008 as an example, the cost of heating this conventional home would rise from $1,600 a year (when oil costs $2 per gallon) to $3,600 per year (at $4.50 per gallon). At this price, heating costs average $300 a month, $50 more than the fixed costs in the net zero house. And the difference will only get bigger as global competition for fossil fuels becomes more intense and prices rise. Does anyone seriously believe fuel oil will still cost $4.50 per gallon in 30 years?

GETTING TO ZERO

If zero energy houses are not in the U.S. housing mainstream yet, there are many signs they will be. The green building boom, higher energy costs, and a lot of publicity about global climate change are likely to motivate buyers who would not have been interested in the past, helping to push net zero into new, non-custom markets. At least eventually.

Marketing the Zero Energy House

Net zero houses have largely been the creatures of the custom home market, in which the high initial cost of renewable energy is less of an obstacle and designers and builders can take more chances. Production builders are not in the best position to gamble with spec houses, especially in a down market, and when they're not familiar with all of the techniques or materials they'll need to master.

Net zero or near net zero subdivisions, however, do exist. In 2001, San Diego-based SheaHomes® began offering energy-efficient houses in its Scripps Highland subdivision. All 306 homes, ranging in cost from $400,000 to $840,000, were sold by November 2003. Of the total, 120 had photovoltaic panels and 293 had solar hot-water systems. The houses were competitively priced and could reduce energy bills by between 30 percent and 50 percent when compared with a conventional house.

According to the National Renewable Energy Laboratory (NREL), the subdivision was the first of its kind from a large production builder in the United States. And buyers liked them. In the first 3½ years after completion, only 5 percent of the houses were resold. In a comparison neighborhood of more conventional homes nearby, 13 percent of the houses were turned over. The high-efficiency houses gained more value than did the conventional houses, and NREL discovered through surveys that homeowners with renewable energy systems were more aware of household energy use and more likely to use less.

Since then, similar subdivisions and demonstration homes have started cropping up, built by companies like Pardee Homes®, Clarum Homes, Morrison Homes, the John Wesley Miller Companies, and Centex®.

Modular net zero

Net zero thinking is also finding its way into modular and panelized housing. In Maine, for example, Kaplan Thompson Architects of Portland collaborated with Bensonwood® Homes and others to produce a small, prefabricated net zero home called BrightBuilt Barn, a panelized structure that can be erected in a couple of days (see the photo on pp.230–231). The 700-sq.-ft. building sells for $165,000 (not including any renewable energy) and comes with R-40 walls, floors, and roof. If that seems a little pricey, Jesse Thompson, one of the firm's principals, says the firm also is working with a modular manufacturer on a 1,680-sq.-ft., three-bedroom house called Modular Zero.

The 1,680-sq.-ft. house in Falmouth, Maine, is the prototype for a modular net zero house under development by Kaplan Thompson Architects and Keiser Homes, a modular manufacturer. With a 6.3kW photovoltaic array and solar hot-water panels, its designers believe the buildings would be net zero in a climate as challenging as northern Maine.

The house will come with triple-glazed windows, R-values of 40 to 60, and a heating load low enough to be met by an inexpensive mini-split heat pump, even in a cold climate. While pricing was still being worked out in mid-2009, Thompson said he had "high hopes for real affordability for the total package."

Marketing strategies

As net zero and green buildings become more common, the real estate industry is looking for ways to market the houses more effectively. The Green Resource Council, part of the National Association of Realtors®, trains agents about green building and has developed a website to make information more readily available to brokers. A few local multiple listing services, which are not directly owned or controlled by the Realtor® association, include information on green features. But council director Al Medina estimates that only 1 percent or less of the nation's listing services make specific provisions for green or net zero features, and that no formal methodology for placing a market value on green or net zero homes has been adopted by the real estate appraisal industry. Although there a few real estate websites that specialize in green buildings (www.listedgreen.com and www.ecobroker.com among them), their listings are not extensive.

In all, a recognized means of separating net zero or even green houses from the rest of the pack seems some distance away. Whether builders will be able to charge a premium for net zero houses and how they will make contact with buyers aren't clear. "I think at the end of the day it's all about marketing and selling," Medina said. "It's so important to educate and properly market a green home. If there is a premium and the developer did spend some extra dollars, there has to be a way to say, 'Here's your payback for those additional dollars.' "

A collaboration between Kaplan Thompson, a Portland, Maine, architectural firm, and Bensonwood Homes, a New Hampshire building company, resulted in BrightBuilt Barn, a 700-sq.-ft. panelized net zero home in Rockport, Maine. Green LED lights around the skirt of the building indicate that the building is producing surplus energy.

A traditionally styled wraparound porch helps the Pill-Maharam home fit in with its Vermont neighbors. A grate in front of the door helps keep sand and mud out of the house.

A House That's Also a Classroom

People with beautiful gardens may put their house on a local garden tour. On a given day, strangers show up to see what's growing and get ideas for their own gardens. So it is with at least some people who live in net zero houses. There's an element of pride in showing off a house with no energy bills, but it's also an opportunity for some gentle proselytizing.

Pill and Maharam put their house on an energy tour shortly after they moved in. "I was amazed at how many people came on that tour," Maharam said. "The first year we had it on the tour close to 80 people came through and we had only been in the house a few months."

While strangers traipsed through the house eyeing appliances, the concrete floor, the modern interior, the picture-postcard views, Pill was in the basement explaining the heat-recovery ventilator, the inverter, and the geothermal heat system. "David never left the basement," Maharam said. "He never left the basement because people just wanted to really understand the systems. They had just specific questions . . . they really wanted to know the specifics."

Far from putting a Private sign up at the end of her driveway and enjoying Squam Lake in peace, Bindley is open and welcoming and eager to share her home and her experience with others. "It was more expensive than I thought it would be, but in terms of evaluating how you spend your money, I think it was a good investment." Why? Because it becomes an opportunity to change the way people look at their houses? "Yes," she said, "and because it's possible to live without fossil fuel."

Case Study

Affordable Housing Meets Zero Energy

Habitat for Humanity and government researchers team up for a demonstration house in a cold climate

In theory, homeowners with deep pockets could meet a zero energy goal simply by adding photovoltaic panels until there was enough electricity to run all the appliances, mechanical equipment, and miscellaneous plug loads in the house. No matter how wasteful the occupants, or inefficient the building, it could eventually be made into a zero energy dwelling. What if the opposite were true? What if every penny in the construction budget counted and the object was energy self-sufficiency at a cost that almost every American homebuyer could afford?

SPECIFICATIONS

House size: 1,284 sq. ft.

Average heating degree days: 5,988

Wall and roof construction: Double 2×4 walls, raised-heel truss roof

Insulation type: Fiberglass batts, blown-in fiberglass

R-values: R-40 walls; R-60 roof; R-30 floor

Windows: Double-glazed, low-e

Photovoltaic capacity: 4kW

Air-conditioning: None

Cost per square foot: $116

Designer: National Renewable Energy Laboratory, Habitat for Humanity

General contractor: Habitat for Humanity, Metro Denver

The 1,280-sq.-ft. Habitat for Humanity house in Wheat Ridge, Colorado, is as unassuming as it is energy efficient. The project showed that net zero performance can be achieved on a modest budget using conventional building practices and simple materials.

This was the premise for a house designed by a team of two Habitat for Humanity managers, two researchers from the National Renewable Energy Laboratory, and two Habitat energy specialists. The house was built in 2005 by Habitat Metro Denver in the suburban community of Wheat Ridge, Colorado. Habitat houses are built by volunteers, often with donated materials and fixtures, and sold to qualifying buyers at the lowest possible cost.

The design team wanted a zero energy building that could be replicated by other Habitat for Humanity projects, that used construction techniques suitable for volunteer labor, and that required no special expertise to operate. It would be an everyman's house. The result is a three-bedroom, 1,284-sq.-ft. house that cost a total of $149,351 (including the land), about $116 per sq. ft. Although the house is unlikely to win any architectural prizes, it is sturdy, comfortable, well ventilated, and, most important, a net producer of energy.

Simple Construction Methods, High Performance

A tight, well-insulated building envelope is a prerequisite for zero energy construction, mainly because the high cost of solar electric power makes an inefficient house impractical and unaffordable. Designers often rely on structural insulated panels or spray-in polyurethane foam, which provide very high R-values and an effective air barrier in one shot. But for Habitat, both of these building components represented the wrong mix: high material costs and low labor costs.

DOUBLE 2×4 WALLS

Structural insulated panels (SIPs) are one way that net zero builders get very high R-values in walls and ceilings. They have a number of advantages, but on the whole they are more expensive than framing a house conventionally. At the Wheat Ridge project, builders used a less-expensive approach: a double 2×4 wall separated by a 3½-in. gap. The assembly left room for three distinct layers of fiberglass batt insulation: one installed vertically in each of the two stud walls and a third where batts could be installed horizontally between them.

Wall R-values are a very respectable 40, and the assembly also minimizes "thermal bridging," the loss of heat directly through framing members in a standard wood-framed wall. The other advantage is familiarity. Framing walls with 2×4s is basic carpentry. It is inexpensive and requires neither specialized tools (such as a crane that's typically used to set SIPs walls) nor tricky techniques, a perfect combination for Habitat projects.

Double-wall 2×4 construction was less complicated and less expensive than using structural insulated panels or spray-in foam insulation. Walls have a nominal R-value of 40.

Designers also considered straw-bale construction, which has an attractive balance of low material costs and high labor requirements. That was rejected because there are no real construction standards for straw-bale houses, and because the team doubted the technique could be easily replicated by Habitat Metro Denver. Ultimately, the design team settled on a simpler and less expensive way that still gives exterior walls a nominal R-value of 40.

The house is 26 ft. by 46 ft., built over a crawl space, with raised-heel trusses, which allow a deep layer of insulation all the way to the eaves. In all, the house got a full 2 ft. of blown-in fiberglass insulation in the ceiling (R-60). Designers used windows that are orientation specific, meaning that windows on the north, east, and west sides are double-glazed low-e with a low solar heat gain coefficient (0.27) and a U-value of 0.22 to control heat losses and heat gain. On the south side, windows

Raised-heel roof trusses allow a thick layer of blown-in insulation in the ceiling to extend all the way to the outside of the wall plate.

Roof overhangs are extra wide, allowing winter sun to reach into the house but keeping heat at bay during the summer. This key element of passive solar design can be an important step in reducing energy costs.

have a high solar heat gain coefficient (0.58) to make the most of the winter sun for space heating. They added 3-ft. roof overhangs to block the summer sun and prevent overheating.

Designers chose a heat-recovery ventilator that supplies fresh air to the living room and bedrooms and exhausts stale air from the kitchen and bathroom. In keeping with Habitat guidelines, there is no mechanical air-conditioning.

Inexpensive Heat System for Smaller Loads

A variety of heating options were considered, including a ground-source heat pump, a natural gas furnace, electrical-resistance baseboard, solar tied to a radiant floor distribution system or baseboard heater, and a solar system that combined space and water heating.

The building's passive solar design convinced designers to stay away from expensive, oversize heating equipment. That included a ground-source heat pump, usually an attractive option because of its efficiency and ability to cool the house in summer. But

Housewrap beneath the exterior siding is sealed at all seams with housewrap tape, part of the strategy for providing a tightly sealed building envelope.

ground-source heat pumps also are expensive, and such a system would need a central distribution system. The house would need so little supplemental heat that this option looked like a waste of money. In fact, solar energy alone was expected to be enough to heat the house on sunny winter days.

In the end, the house got a hybrid heating system that combines a direct-vent natural gas furnace in the living/dining area with 750w electric baseboard heaters in the bedrooms. Each appliance has its own thermostat, and the equipment is both inexpensive and simple.

The design team weighed the inherent disadvantages of using a fossil fuel to heat the house: Not only is there a fixed monthly fee for having the hookup, but burning natural gas also produces greenhouse gases that contribute to global warming. Designers also knew a photovoltaic system could produce enough excess power and export it to the utility's grid to offset the energy represented by the natural

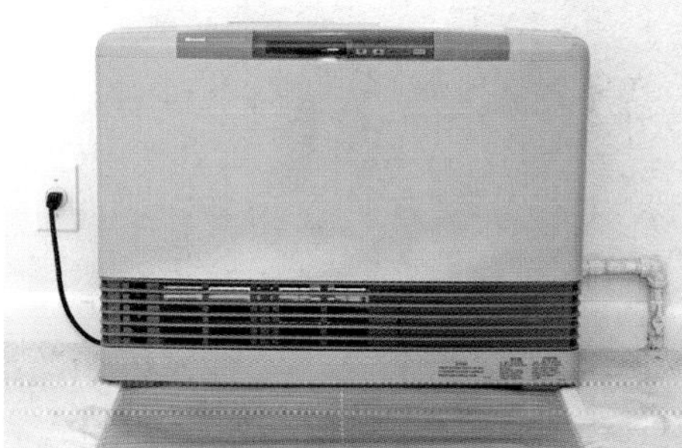

Space heating is provided by this gas-burning appliance in the living room along with electrical resistance heat in the bedrooms. There is no mechanical heat distribution system.

SOURCE ENERGY VS. SITE ENERGY

There's more than one way to calculate net energy consumption. It may seem a little convoluted at first, but comparing "site" energy to "source" energy really does make sense. Just work through the numbers.

Considered on a site basis, the Wheat Ridge house is a net consumer of energy. The combined consumption of electricity and natural gas (the equivalent of 5,250kWh) slightly outweighed the production of power from the photovoltaic panels (5,127kWh).

But a single kilowatt hour of PV-generated electricity is worth more than 1kWh that comes from the utility's grid. Why? Because producing electricity at a big power plant and then transmitting it to the end user is inherently inefficient. In fact, for every kilowatt hour of power produced at the house, 3.16kWh of electricity is saved at the source.

With that multiplier in mind, the 5,127kWh of electricity made at the Wheat Ridge house is the equivalent of 16,201kWh of power produced by the utility. Because the house used a total of 13,205kWh of source electricity, it actually showed a net energy offset of 3,176kWh. Expressed a little differently, the house produced 124 percent of the energy it consumed on a source basis.

gas the house consumed. This approach balances "source" energy rather than "site" energy to get to net zero.

Domestic hot water comes from a 96-sq.-ft. solar collector mounted on the roof and a 200-gal. storage tank, a system that promised to save 88 percent of the energy that would be used to heat water in a conventional Habitat house that was being built nearby. Designers found they could mount the collectors at the same angle as the roof pitch, thereby making them less obtrusive, and lose very little solar potential. A tankless natural gas heater is the backup to the solar collectors.

Photovoltaic Array for a "Typical" Family

Designers reduced electrical loads by specifying EnergyStar appliances and compact fluorescent light bulbs, but the balance of electricity consumption—all of the small appliances, electronics, and other gizmos that occupants would plug in—was largely outside their control. This category, called miscellaneous electric loads (MELs), is expected to consume 57 percent of all electricity in the zero energy Habitat home; a comparison Habitat house was expected to use 28 percent.

A tankless hot-water heater in the mechanical room (below) acts as the backup for the solar hot-water collectors on the roof.

Photovoltaic panels were attached to the roof conventionally, with brackets that are fastened through the shingles. The low angle of the array makes the renewable energy system unobtrusive, virtually invisible from the street.

The National Renewable Energy Laboratory (NREL) says these loads are "highly unpredictable" and may vary considerably, depending on how a family lives. But assuming that people who used typical amounts of electricity would occupy the house, the designers called for a 4kW system. If weather and household patterns were as expected, energy use should net out at zero. But change those parameters, and the house could be either a net producer or consumer of electricity.

The solar electric system would have cost $40,000 at full retail but was discounted for the Habitat program to $17,489. After rebates, the system cost Habitat nothing. In total, features that allowed the house to be a zero energy building added about $23,400 in construction costs when compared to a standard-practice home.

Energy Performance Beats Predictions

So how did it work out? Better than expected. Over a 12-month period, the house and its occupants used a total of 3,585kWh of electricity while generating 5,127kWh. Natural gas

HOMEOWNERS HAVE MORE CONTROL

In a typical Habitat house, 28 percent of the power consumed goes to appliances and other devices that plug into the wall. In the net zero house at Wheat Ridge, Colorado, the electricity needed for lighting, space heating, and water heating is substantially lower, meaning that a larger proportion of energy consumption is under the direct control of homeowners. More control equals greater potential savings.

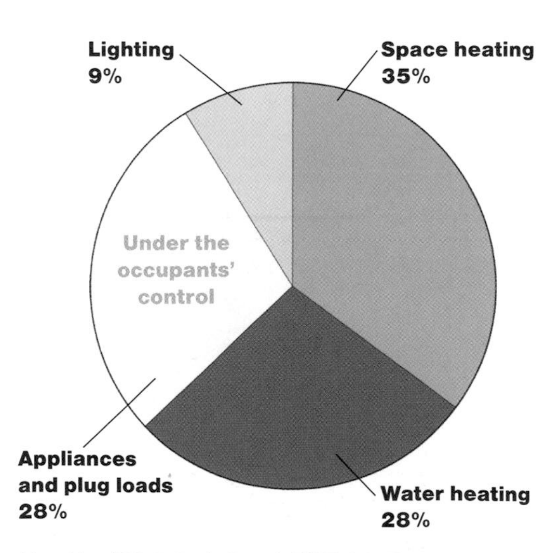

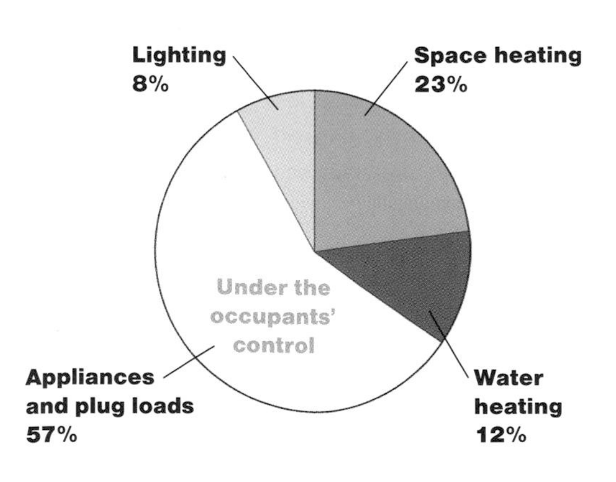

Adapted from NREL/Habitat for Humanity ZEH Technical Report

consumption was the equivalent of 1,665kWh of electricity. When comparing total source energy, the house used 13,025kWh and produced an offset of 16,201kWh (124 percent of what was consumed) for a net energy offset of 3,176kWh. Because of utility rate structures and fixed utility charges, the homeowner paid an average monthly utility bill of $17. If homeowners could have taken advantage of feed-in tariffs (see p. 107), that bill for electricity could easily have been a payment.

Even while the PV system was topping expectations, the solar hot-water collector and heat distribution didn't go exactly as planned. The design team had predicted the homeowners would use about 63 gal. of hot water a day, when in reality they used just about 20 gal. The numbers made the design team wonder whether it would be cheaper to replace the solar hot-water system, which cost about $7,000, with a more robust photovoltaic array. In this instance, given Colorado's gener-

THE UPS AND DOWNS OF SOLAR ELECTRICITY

The production of electricity from the 4kW array at Wheat Ridge, Colorado, follows a predictable track: lower output during bad weather and steady generation throughout milder months. Even with heavy snow during the winter of 2006–2007, the house was a net producer of electricity.

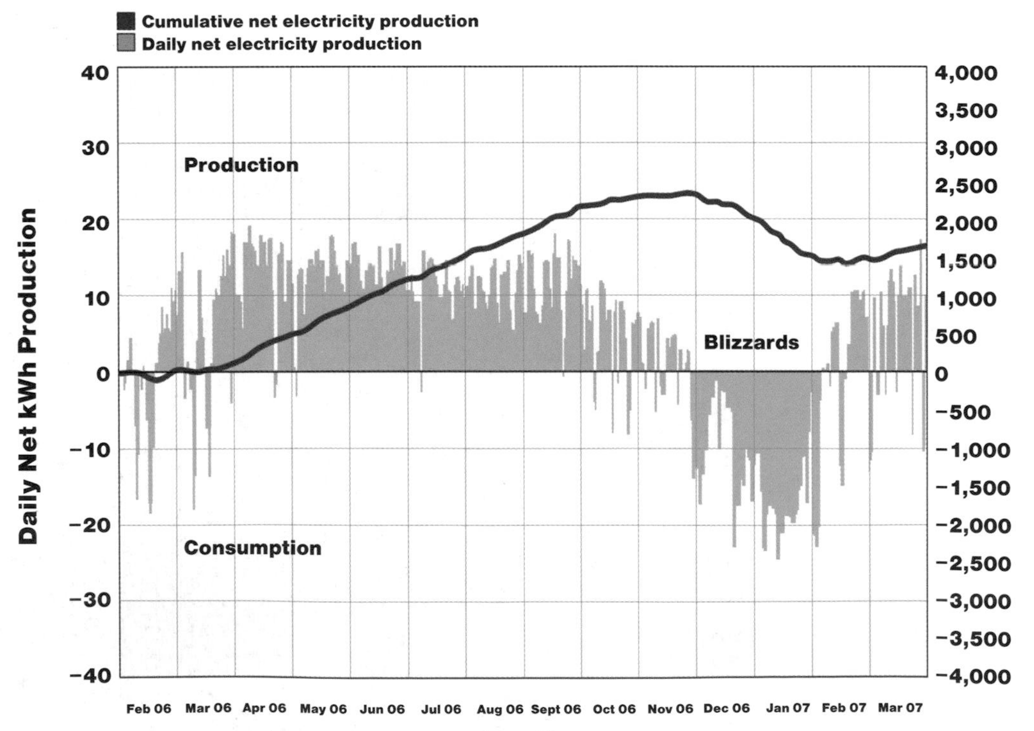

Adapted from NREL/Habitat for Humanity ZEH Technical Report

ous rebates for PV, it would have cost about $4,000 less to incrementally increase PV capacity and install an electric tank water heater.

On the heating side, the design team expected the direct-vent gas heater would do the lion's share of the work. But the electric baseboard heaters actually consumed 60 percent of total space heating site energy. It told designers that heat distribution to the bedrooms was inadequate and that if natural gas were to become a bigger contributor then another heater or a heat distribution system would have to be added.

A 4kW photoelectric array produced more electric energy than the family consumed. When the electricity equivalent of the gas burned in the space heater was added in, the house was still a net producer of power on a source basis.

A happy day: Dedication ceremonies for the Habitat project in Wheat Ridge welcomed new owners to a home that is comfortable, well ventilated, and energy efficient.

APPENDIX 1

ENERGY MODELING SOFTWARE

When designing a net zero energy home, small changes can make a tremendous difference and optimizing the design is key. Factors such as window size, window placement, and insulation choices cannot be left to guesswork. Energy modeling software can help guide a designer to choose the right combination of elements by projecting the results of specific changes long before the house is ever built. There are many different types of energy modeling software available, and each has its own pros and cons. It's useful to try more than one software program to find the product that best suits your needs. The following is an alphabetical list of some of the programs available. The U.S. Department of Energy maintains a list on its website (http://apps1.eere.energy.gov/buildings/tools_directory/alpha_list.cfm).

BESTEST

The Building Energy Simulation Test (BESTEST) is a program for testing and diagnosing the simulation capabilities of the building envelope portions of energy simulation programs. Three versions are available: HERS BESTEST (emphasis on modeling houses), IEA BESTEST (detailed hourly simulation programs), and Florida BESTEST (for hot-humid climates).

CLIMATE 1

Global climate database with more than1,200 station data sets. Climate data are shown graphically and in lists and include mean, maximum, and minimum temperature; relative humidity; precipitation; sunshine hours; and wind speed and direction.

DesignBuilder

Provides a range of environmental performance data such as energy consumption, internal comfort, and HVAC component sizes.

DOE-2

Calculates energy performance and life-cycle cost of operation; can be used to analyze energy efficiency of given designs or efficiency of new technologies. Can also be used for utility demand-side management and rebate programs.

ECOTECT

Environmental design tool that couples three-dimensional modeling with solar, thermal, lighting, acoustic, and cost analysis functions.

Energy-10

Conceptual design tool that performs hourly whole-building energy analysis, including dynamic thermal and daylighting calculations. Especially effective at facilitating the evaluation of energy-efficient building features in the early stages of the design process.

EZDOE

Calculates the hourly energy use of a building and its life-cycle cost of operation using information on the building's location, construction, operation, and heating and air-conditioning systems. Takes into account complex thermal storage effects of various building materials and accurately simulates the operation of all types of heating and cooling plants.

PHPP

Developed in Germany in conjunction with the PassivHaus movement, PHPP has excellent passive solar support and gives monthly heating and cooling data. Has a useful passive cooling calculator and determines losses for hot-water distribution, ventilation ducting, and more.

REM/Design

Developed specifically with the needs of homebuilders, remodelers, energy consultants, and designers in mind to calculate heating, cooling, hot water, lights, and appliance loads as well as consumption and costs for single-family and multifamily designs in over 250 North American cities.

WATERGY

Screens sites for potential water conservation opportunities and illustrates the energy savings that result from water conservation activities.

APPENDIX 2

AIR CHANGES PER HOUR: 50 PASCALS OF PRESSURE

Air movement in and out of a building through penetrations in a building's envelope changes dramatically depending on whether wind is blowing and whether a fan is on in the ventilation system. Deriving the natural air changes per hour (NACH) in the building depends on the interaction of the stack effect and wind along with the shape of the house. It is a dynamic condition that is always in flux, depending on the weather outside.

The standard developed in Canada and Sweden was 50 pascals, which was later adopted by the American Society of Heating, Refrigeration, and Air-Conditioning Engineers (ASHRAE). It is the negative pressure created by a blower-door test inside a home to simulate air leakage under wind pressure. Smoke tests reveal that 10 pascals is roughly equivalent to a wind of 9 mph to 13 mph. Standard atmospheric pressure at sea level is 101,325 pascals, so a difference of 50 pascals will not be discernible to someone standing in the room.

A blower-door test is like a snapshot vs. a movie. The results of the "snapshot" ACH at 50 pascals are related to the "movie" NACH (actual air changes per hour) by the following rule of thumb:

ACH at 50 Pascals ÷ 20 = NACH

If you measured 3,200 cfm at 50 pascals, and you have a two-story, 2,000-sq.-ft. house with 8-ft. ceilings, then:

Home air volume = 2,000 sq. ft. × 8 ft. = 16,000 cu. ft.

ACH at 50 pascals = (3,200 cfm × 60 min) ÷ 16,000 cu. ft. = 12 ACH50 = 12 ÷ 20 = 0.6 NACH

The best ACH algorithms claim accuracy only within 25 percent.

For an online tool to estimate air changes, go to www-epb.lbl.gov/ventilation/program.html

The PassivHaus maximum air leakage rate is 0.6 ACH50 (about 0.04 NACH). The ASHRAE 62.2 standard requires mechanical ventilation if the house is tighter than 0.35 NACH (ACH50=7). Typical production-built homes (even under modern building codes) will score at about 0.5 NACH or higher. Adam Stenftenagel of Sustainably Built in Boulder, Colorado, reports that the average for well-built, near zero energy new homes is between 0.1 NACH and 0.2 NACH.

RESOURCES

Advanced Energy
909 Capability Dr., Suite 2100
Raleigh, NC 27606
919-857-9000
www.Advancedenergy.org

Affordable Comfort Inc
32 Church St., Suite 204
Waynesburg, PA 15370
800-344-4866
www.Aci.org

American Council for an Energy-Efficient Economy
529 14th St. N.W., Suite 600
Washington, DC 20045-1000
202-507-4000
www.aceee.org

Architecture 2030
607 Cerrillos Rd.
Santa Fe, NM 87505
info@architecture2030.org

Build It Green
1434 University Ave.
Berkeley, CA 94702
510-845-0472
www.builditgreen.org

Building Knowledge Inc.
541 West 98th St.
PMB 213
Minneapolis, MN 55420
952-944-5605
www.buildingknowledge.com

Building Performance Institute (BPI)
107 Hermes Rd., Suite 110
Malta, NY 12020
877-274-1274
www.bpi.org

Building Science Corporation
30 Forest St.
Somerville, MA 02143
www.buildingscience.com

California Energy Commission
Media and Public Communications Office
1516 Ninth St., MN-29
Sacramento, CA 95814-5512
916-654-4287
www.energy.ca.gov

Canada Mortgage and Housing Corporation (CMHC)
700 Montreal Rd.
Ottawa, Ontario
K1A 0P7
613-748-2000
www.cmhc-schl.gc.ca/en

Energy and Environmental Building Alliance
6520 Edenvale Boulevard, Suite 112
Eden Prairie, MN 55346
952-881-1098
www.eeba.org

Energy Information Administration
1000 Independence Ave., SW
Washington, DC 20585
202-586-8800
www.eia.doe.gov

Federal Emergency Management Agency (FEMA)
500 C St. S.W.
Washington, DC 20472
202-646-2500
www.fema.gov

Florida Solar Energy Center
1679 Clearlake Rd.
Cocoa, FL 32922
321-638-1000
www.fsec.ucf.edu

Forest Stewardship Council
212 Third Ave. North, Suite 280
Minneapolis, MN 55401
612-353-4511
www.fscus.org

GreenBuilding.com
57 Acorn La.
Boulder CO 80304
303-444-7044
www.greenbuilding.com

National Association of Home Builders
1201 15th St., NW
Washington, DC 20005.
800-368-5242
www.nahb.org

NAHB Research Center
400 Prince George's Blvd.
Upper Marlboro, MD 20774
800-638-8556
www.Toolbase.org

National Institute of Building Sciences
1090 Vermont Ave., NW, Suite 700
Washington, DC 20005
202-289-7800
www.nibs.org

National Renewable Energy Laboratory
1617 Cole Blvd.
Golden, CO 80401
303-275-3000
www.nrel.gov

Net Zero Home Energy Home Coalition (Canada)
613-823-8079
http://www.netzeroenergyhome.ca/

Northeast Sustainable Energy Association
50 Miles St.
Greenfield, MA 01301
413-774-6051
www.nesea.org

Oak Ridge National Laboratory
P.O. Box 2008
Oak Ridge, TN 37831
865-574-4160
www.ornl.gov

Passive House Institute (U.S.)
110 S. Race St., Suite 202
Urbana, IL 61801
217-344-1294
www.passivehouse.us

Structural Insulated Panel Association
P.O. Box 1699
Gig Harbor, WA 98335
253-858-7472
www.sipa.org

U.S. Department of Energy
Energy Efficiency & Renewable Energy
1000 Independence Ave., SW
Washington, DC 20585
877-337-3463
www1.eere.energy.gov

CASE STUDY RESOURCES

High Heat (Lakeland, Florida)
Designer: Florida Solar Energy Center
(www.fsec.ucf.edu)
General contractor: Strawbridge Construction

A Butterfly Home in the City (Chicago)
Designer: Farr Associates (www.farrside.com)
Energy modeling: dbHMS (www.dbhms.com)
General contractor: Goldberg General Contracting
(www.ggcinc.net)

Outside-In Retrofit (Somerville, Massachusetts)
Designer: Steve Baczek Architect
General contractor: Byggmeister Design Build
(www.byggmeister.com)
HERS Rater: Mike Duclos, Energy Efficiency Associates
(www.eeassociates.com)

PassivHaus Retrofit (Berkeley, California)
Designer: Nabih Tahan
(www.nabihtahanarchitect.com)
General contractor: Christopher Polk
Home performance contractor: George Nesbitt, Environmental Design/Build
(www.houseisasystem.com)

Bringing 1887 into the 21st Century (Santa Barbara, California)
Designer: Thompson Naylor Architects
(www.thompsonnaylor.com)
General contractor: Allen Associates
(www.dennisallenassociates.com)

Target Zero House in Taos (Taos, New Mexico)
Designer: Joaquin Karcher, Zero E Design
(www.zeroEdesign.com)

The Good Life in Vermont (Charlotte, Vermont)
Designer: David Pill and Hillary Maharam, Pill-Maharam Architects
(www.pillmaharam.com)
Energy consultant: Andy Shapiro, Energy Balance Inc.
General contractor: Jim Huntington, New England Housewrights Ltd.

The Next West House (Boulder, Colorado)
Designer: Jim Logan Architects
(www.jlogan.com)
General contractor: Hughes Construction
(www.hughesconstruct.com)

High-End Lakeside Retrofit (Squam Lake, New Hampshire)
Designer: Ben Southworth, Garland Mill Timberframes
(www.garlandmill.com)
Builder: Garland Mill Timberframes

A Solar Jewel in Boulder (Boulder, Colorado)
Designer: Michael Kracauer, architropic
(www.architropic.us)
General contractor: Morningstar Homes
(www.morningstarbuilt.com)

Reinventing the High Rise (New York City)
Designer: Chris Benedict, Architect
Mechanical Designer: Henry Gifford, Architecture and Energy Limited
(www.architectureandenergylimited.com)

All-Electric House: $1.16 a Day (Lenoir City, Tennessee)
Designer: Oak Ridge National Laboratory (www.ornl.gov), Build America
(www1.eere.energy.gov/buildings/building_america)
General contractor: Habitat for Humanity
(www.habitat.org)

Affordable Housing Meets Zero Energy (Wheat Ridge, Colorado)
Designer: National Renewable Energy Laboratory (www.nrel.gov), Habitat for Humanity (www.habitat.org)
General contractor: Habitat for Humanity, Metro Denver

CREDITS

p. ii: © Scott Gibson
p. 2: © Michael Kracauer
p. 4: (left) © Scott Gibson; (right) © David Pill
p. 5: © Scott Gibson
p. 6: Courtesy Cador Pricejones
p. 7: Courtesy Florida Solar Energy Center
p. 8: (top) Courtesy Oak Ridge National Laboratory; (bottom) Courtesy Farr Associates
p. 9: Courtesy Paul Norton, National Renewable Energy Laboratory

Chapter 1

p. 11: © Michael Kracauer
p. 12: © Russell Hamlet, Architect
p. 13: (top) © iStockphoto.com/Petrea Alexandru; (bottom) © iStockphoto.com/John Clines
p. 14: © iStockphoto.com/Soubrette
p. 15: © iStockphoto.com/Chris Price
p. 16: © iStockphoto.com/Ilya Shulman
p. 17: (top) © Allen Associates; (bottom) Marie-France Cloutier
p. 19: © Nancy Schultz
p. 21: © What's Working
p. 23: Courtesy Airia Brands Inc. (Lifebreath)
p. 24: © Jeff Boxer 2009
p. 27: (top) © What's Working; (bottom) © Russell Hamlet, Architect
p. 28: © What's Working
p. 29: (left) © Russell Hamlet, Architect; (right) Paul Norton, Courtesy National Renewable Energy Laboratory
p. 30: (top) © Joaquin Karcher, Zero E Design; (bottom) © What's Working
p. 32: (top) © What's Working; (bottom) © Allen Associates
p. 34: © Russell Hamlet, Architect
p. 37: Photos © What's Working
p. 39: © What's Working
p. 40: (top) © Russell Hamlet, Architect; (bottom) © What's Working
p. 41: Sustainably Built, LLC
p. 43-49: Courtesy Florida Solar Energy Center
pp. 51-57: Photos © Christopher Barrett Photographer
pp. 59-63: Photos © Cador Pricejones

Chapter 2

p. 65: (top) © iStockphoto.com/Sadik Gulec; (bottom) © iStockphoto.com/Jeremy Edwards
p. 67: © David Johnston
p. 68: (top) © iStockphoto.com/Robert Simon; (bottom) © Deirdre Damron
p. 69: © David Johnston
p. 70: © iStockphoto.com/Marc Fischer
p. 72: © Barrett Studio Architects
p. 74: Courtesy National Renewable Energy Laboratory
p. 75: (left) © What's Working; (right) © iStockphoto.com/sandramo
p. 76: Photos © What's Working
p. 77: © iStockphoto.com/Timothy Large
pp. 79-81: Photos © What's Working
p. 83: © Dennis Kim
p. 84: (left) © Nabih Tahan; (right top) © Dennis Kim; (right bottom) © George J. Nesbitt Environmental Design/Build
p. 85: (left) © Nabih Tahan; (right) © Rob Harold
p. 86: © Nabih Tahan
p. 87: © Dennis Kim
pp. 89-94: Photos courtesy of Allen Associates
pp. 97-100: © Joaquin Karcher, Zero E Design
p. 101: © Jon Wood

Chapter 3

p. 103: © Warren Gretz, courtesy National Renewable Energy Laboratory
p. 104: Courtesy Namaste Solar
p. 105: (left) © MicroLink Devices, courtesy National Renewable Energy Laboratory; (right) © Susannah Pedigo, courtesy National Renewable Energy Laboratory
p. 106: © Scott Gibson
p. 109: Courtesy Namaste Solar
p. 110: (left) Courtesy United Solar; (right) © Scott Gibson
p. 111: (left) © Warren Gretz, courtesy National Renewable Energy Laboratory; (right) Courtesy Namaste Solar

p. 112: (top left) Courtesy National Renewable Energy Laboratory; (left middle & bottom) Courtesy Namaste Solar; (right) © Warren Gretz, courtesy National Renewable Energy Laboratory
p. 113: (left) Courtesy Atlantis Energy Systems, Inc.; (right) Courtesy National Renewable Energy Laboratory
p. 114: (top & bottom left) © Rich Small
p. 115: Courtesy National Renewable Energy Laboratory
p. 117: Photos courtesy First Solar
p. 119: (top) © iStockphoto.com/David Sucsy; (middle & bottom) © Gamesa, courtesy National Renewable Energy Laboratory
p. 120: © Scott Gibson
p. 121: © David Pill
p. 122: (left) © Jim Tracy, courtesy National Renewable Energy Laboratory; (right) Appalachian State University, courtesy National Renewable Energy Laboratory
p. 127: © Scott Gibson
p. 128: Courtesy Kingspan Solar
p. 129: Courtesy Rheem Manufacturing Company
p. 132: (left) © Scott Gibson; (right) Courtesy Velux
p. 135: © iStockphoto.com/MorganLane Studios
p. 136: © iStockphoto.com/srbeckle
p. 137: © Warren Gretz, courtesy National Renewable Energy Laboratory
p. 139: © Jim Westphalen
p. 140: Photos © David Pill
p. 142: © Scott Gibson
p. 143-144:© Jim Westphalen
p. 145: © David Pill
p. 147: © Emily Minton Redfield
pp. 148-150: Photos © What's Working
p. 151: (left) © What's Working; (right) © Emily Minton Redfield
p. 152: © What's Working
p. 153: © Emily Minton Redfield
p. 155: (top) © Fletcher Manly, courtesy Garland Mill; (bottom) Courtesy Garland Mill
pp. 156-159: Courtesy Garland Mill
p. 160:© Fletcher Manly, courtesy Garland Mill
p. 161: Courtesy Garland Mill
p. 162: © Terry Dupuis
p. 163: Courtesy Garland Mill

Chapter 4

p. 165: © David Pill
p. 166: © Jim Westphalen
p. 167: © Scott Gibson
p. 171: (top) Scott Gibson, courtesy *Fine Homebuilding* magazine, © The Taunton Press, Inc.; (bottom) © Scott Gibson
p. 174: © iStockphoto.com/Greg Nicholas
p. 175: Photos courtesy Hallowell International
p. 177: © Cadet Manufacturing, 2009
p. 179: Photos courtesy of Nuheat Floor Heating Systems
p. 180: (left) © iStockphoto.com/Joel Johndro; (right) © iStockphoto.com/Maurice van der Velden
p. 181: Courtesy Ecotek
p. 182: © Scott Gibson
p. 183: Courtesy Carrier
p. 184: (left) © What's Working; (right) © iStockphoto.com/Vasko Miokovic
p. 185: (left) © Jim Westphalen; (right) ©Sean Groom, courtesy *Fine Homebuilding* magazine © The Taunton Press, Inc.
p. 188: (left) Courtesy Sanyo; (right) Courtesy Mitsubishi Electric Cooling and Heating Solutions
p. 189: (top) © iStockphoto.com/Stanislav Fadyukhin
p. 191: Courtesy Fantech
p. 192: (left) Courtesy National Renewable Energy Laboratory; (right) © iStockphoto.com/Lisa Fx Photographic Designs
p. 193: Courtesy AirGenerate
p. 195-200: © Michael Kracauer
p. 201: © Photos Allison M. Fleetwood, Jr., Architectural Photography
pp. 203-208: © Photos Henry Gifford
pp. 211-217: Photos courtesy Oak Ridge National Laboratory

Chapter 5

pp. 219-226: Photos © Scott Gibson
p. 227: Courtesy National Renewable Energy Laboratory
p. 229: © Jamie Broadbent, Kaplan Thompson Architects
p. 231: © Trent Bell
p. 232: © Scott Gibson
pp. 234-241: Photos courtesy Paul Norton, National Renewable Energy Laboratory

INDEX

H

I

L

M

N

P

6/2010